KT-441-542

EAST SUSSEX COUNTY COUNCIL
WITHDRAWN
0 7 JUN 2024
19

04404084

MAPPING THE HEAVENS

MAPPING THE HEAVENS

THE RADICAL SCIENTIFIC IDEAS THAT REVEAL THE COSMOS

PRIYAMVADA NATARAJAN

Yale UNIVERSITY PRESS
New Haven & London

Published with assistance from the foundation established in memory of Amasa Stone Mather of the Class of 1907, Yale College.

Yale University Press books may be purchased in quantity for educational, business, or promotional use. For information, please e-mail sales.press@yale.edu (U.S. office) or sales@yaleup.co.uk (U.K. office).

Designed by James J. Johnson.
Set in Adobe Caslon Pro and Whitney types by Integrated Publishing Solutions.
Printed in the United States of America.

ISBN 978-0-300-20441-4 (cloth : alk. paper)

Library of Congress Control Number: 2015953466
A catalogue record for this book is available from the British Library.

This paper meets the requirements of ANSI/NISO Z39.48–1992 (Permanence of Paper).

10 9 8 7 6 5 4 3 2 1

To Amma and Appa

CONTENTS

PREFACE

Our map of the cosmos has altered dramatically in the past hundred years. In 1914, our own galaxy, the Milky Way, constituted the entire universe—alone, stagnant, and small. Cosmological research still relied fundamentally on classical conceptions of gravity developed in the seventeenth century. Modern physics and the triumphs of general relativity have shifted humanity's entire comprehension of space and time. Now we see the universe as a dynamic place, expanding at an accelerating rate, whose principal mysterious constituents, dark matter and dark energy, are unseen. The remainder, all the elements in the periodic table, the matter that constitutes stars and us, contributes a mere 4 percent of the total inventory of the universe. We have confirmed the existence of planets orbiting other stars. We question the existence of other universes. This is remarkable scientific progress.

Cosmology, perhaps more essentially than any other scientific discipline, has transformed not only our conception of the universe but also our place in it. This need to locate ourselves and explain natural phenomena seems primordial. Ancient creation myths shared striking similarities across cultures and helped humans deal with the uncertainty of violent natural phenomena. These supernatural explanations evoke a belief in an invisible and yet more powerful reality, and besides, they rely deeply on channeling our sense of wonder at the natural world. The complex human imagination enabled ancient civilizations to envision entities that were not immediately present but still felt real. Take for instance Enki, the Sumerian god of water whose wrath unleashed floods, or the Hindu god of rain and thunderstorms, Indra, whose bow was the rainbow stretched across the sky with a lightning bolt as his arrow. The most powerful myths are the ones that force us to take huge leaps of imagination but, at the same time, help us to remain rooted.

As a child growing up in India, I also felt this drive to locate myself in the world. My first guide was the *Encyclopaedia Britannica.* Thirty-two volumes of the fifteenth edition, sitting on my parents' bookshelf, represented for me everything that was known at that time. Enchanted, I immersed myself in ancient maps, maps that guided the voyages of exploration, and maps of the sky. The stars transfixed me. My personal cartographic adventure also gave me my first taste of scientific research. Programming a Commodore 64, I wrote code to generate the monthly sky map over Delhi for a national newspaper. Thus began my love affair with the idea of discovery and exploration. I studied physics, mathematics, and philosophy during my undergraduate years at the Massachusetts Institute of Technology. My curiosity next led me to graduate study in MIT's Program in Science, Technology, and Society, then across the pond to Cambridge University for a PhD in astrophysics. Now, as an active scientist, I continually draw on my intellectual training in the history and philosophy of science to reflect more deeply on the process of scientific discovery and how it shapes the knowledge we produce.

At its heart, my research as a theoretical astrophysicist, mapping dark matter and understanding the formation of black holes, is driven by the same sense of wonder and search for explanation of the universe that the ancients probably felt. I am still engaged in exploring the meanings of maps and how they anchor us, matters that first intrigued me as a girl in Delhi. My work exploits the bending of light from distant galaxies, gravitational lensing, to map the invisible dark matter that causes these deflections. I also investigate the formation and growth of the universe's most bizarre and enigmatic objects, black holes. Currently, I am involved in one of the largest and most innovative mapping exercises of the universe ever undertaken: the Hubble Frontier Fields Initiative. The goal of this project is to peer more deeply into the distant universe and to map dark matter more accurately than ever before. Between 2014 and 2017, a significant portion of the observing time of the cameras aboard the Hubble Space Telescope will be devoted to this enterprise. Of course, I am one of many researchers contributing to the greater map of the universe that these unique data will provide. Many new, exciting discoveries lie ahead. We, like the generations of scientists who came before us, may find ourselves challenged to completely rethink the status quo.

While there are many books that tell the history of cosmological discoveries, my goal here is to recount how scientific ideas have been developed, tested, debated, and eventually accepted. You certainly need not be an astrophysicist to follow the story, and the examples that I chronicle, though cosmological, are meant to illustrate much broader trends in scientific research and discovery. In particular, I trace the development of radical scientific ideas that have continually reshaped our cosmic map. I find the process by which these ideas have gained traction and advanced from obscurity to acceptance deeply fascinating. In cosmology, the making and remaking of maps often reflects this process, leaving behind cartographic evidence. Seismic shifts in our view of the universe have required overhaul of our knowledge maps in the past century. But the acceptance of new ideas is not linear or instantaneous and is always contested. As scientists have challenged prevailing understandings of the universe, our world view and metaphorical map have morphed ceaselessly, requiring us to adapt and be open to change.

This is a story of extraordinary leaps of imagination, of radical new ideas fueled by discoveries and data. The journey to acceptance of an idea reveals many other facets of science—the emotional, psychological, personal, and social dimensions that extend beyond the purely intellectual pursuit of knowledge. This view is contrary to the popular perception of unbiased inquiry by purely objective researchers engaged in deriving fixed truths from nature. The fact is, science is ultimately a human endeavor; therefore, it is laced with subjectivity.

Controversies and disagreements within the scientific community are an integral part of the pursuit, and these debates are illuminating precisely because they show us—in high relief—how new ideas struggle to finally garner acceptance. To that end, I examine why disputes arise among the community of cosmologists and how they resolve. Such disputes have not ceased, and this ongoing engagement is inherent to the provisional nature of science. The scientific mind is honed during training to be nimble, and the practice of science tests this agility on a daily basis. This inoculates scientists against disorienting shocks when a preponderance of new data and evidence changes the best current understanding. I show how cosmologists have coped with these frequent shifts and reconfigured their knowledge maps by creatively harnessing the power of curiosity and wonder.

It is the powerful confluence of new ideas and new instruments that has transformed our knowledge of the cosmos. Take for instance the invention of the spectrograph, which separates light into its component frequencies, allowing remote study of the chemical composition of distant stars; powerful telescopes and sensitive cameras that produce incredibly high-resolution images; or computers that can store and process vast amounts of data—these have all sparked the generation of new ideas and enabled scientists to test their validity.

In the past few decades, researchers have probed farther into space and back in time with sophisticated satellites and detectors. We have described relics, in the form of electromagnetic radiation, that have brought us tantalizingly close to the moment of creation—the big bang. And in our own backyard we have discovered more than a thousand planets orbiting nearby stars outside our solar system. Yet mysteries still abound.

In the majesty of the night sky, we once derived comfort from fixed stars—points of light that, since antiquity, could be relied on to rise and fall predictably. In 1718, the British astronomer Edmund Halley, the second-ever appointed astronomer royal of Britain, found that these stars in fact moved and their positions changed over time. For example, the stars Sirius, Arcturus, and Aldebaran had strayed far from their positions as chronicled by the ancient Greek astronomer Hipparchus about two thousand years prior. The fixed stars apparently wandered.

Such disorienting discoveries are common in cosmology, and our current understanding of the expanding and accelerating universe has similarly upended our sense of stasis. It all began in 1543 when Nicolaus Copernicus shifted the pivot from the earth to the sun, permanently altering our place in what today we call the solar system but at the time constituted the entire cosmos. Unfixing the stars led to greater changes. In the 1920s, first with the discovery of other, distant galaxies, proving that the Milky Way was merely one among many, and then with evidence that the cosmos was expanding, the astronomer Edwin Hubble set the entire universe adrift. Today we have images and data of several million galaxies, many of them so distant that the light we see originated from them when the universe was in its infancy, a mere billion years old, barely a fraction of its current age of 13.8 billion years. Such stories are part of a larger tale of how we arrived at some of the most remarkable ideas in cosmology in the past hundred years and how those ideas gained traction. The human side of science, rife with personal rivalries, clashes of ambitions, and the search for fame, has both hindered and propelled many a discovery. The human desire for security and the preservation of the status quo kicks into gear when we are confronted with dramatic change. This instinct for stasis colors our reactions to radical new ideas and impedes acceptance of revisions to our deeply held world view. Scientists are not exempt from this and often resist change until convincing evidence persuades them.

The notion of a clockwork universe governed by universal laws such as Isaac Newton's concept of gravitation was accepted rapidly because the picture reinforced a stable and steady universe. Newton's discoveries, novel as they were, rooted us more firmly and provided a sense of being fixed. Even Copernicus's revolutionary discovery of the heliocentric universe, while it was opposed rather famously in some quarters, was in the end accepted widely, as it retained the fixed notion of our universe and merely rearranged the focal point from us to the sun at the center.

The great disruptive cosmological discoveries of the twentieth and twenty-first centuries include the expanding universe, dark matter, black holes, the big bang model, the accelerating universe, and numerous planets and planetary systems around other stars—discoveries that have opened the door to an ever-shifting cosmos in flux, where we are simultaneously unique and yet insignificant in many ways in the grand scheme.

I trace the passage of these deeply disorienting ideas from conception to acceptance, highlighting their twists and turns and cataloguing their indelible and transformative impact on our ever-evolving world view. These revolutionary shifts from a fixed, static universe to one that is completely unfixed have required continual overhaul and remaking of our cosmic view. By their nature, these advances in cosmology leave us unmoored. Such reframing scientific discoveries, deliberate or serendipitous, often cause discomfort even for the discoverer. How scientists grow to accept new ideas and rewrite their knowledge maps not only reveals how science works but also provides insights into what catalyzes these shifts in belief. The sanitized account of science as an objective method to derive eternal truths from nature writes out the emotions and passions that drive us scientists. The inherently provisional nature of this pursuit is best illustrated by the fact that scientific progress occurs in fits and starts, leading to unanticipated and initially unfathomable places. I unpack this complicated, exhilarating process in light of the changing practice of science. We are now in the era of big science, which represents huge investments of human intellectual capital and other resources, functioning with large teams and the expertise of many technically skilled investigators. This shift in the scale of the research endeavor has transformed how all scientists, including cosmologists, work.

The Sloan Digital Sky Survey collaboration, whose goal was to provide detailed three-dimensional maps of a third of the entire sky, for instance, relied on a team of several hundred scientists drawn from more than forty research institutions around the globe. Although research collaborations in cosmology are not as large as those in experimental particle physics, where participants run into the thousands, astronomy has witnessed a dramatic shift from even just thirty years ago, when it was not uncommon to work in groups of two or three. As cosmology has matured, driven by the use of ever more sophisticated instruments and technologies, both scientists and the work require more resources. This dramatic change in the mode of research and the complexity of instruments deployed has also spawned new interdisciplinary fields, for example astroparticle physics, at the boundary of astrophysics and particle physics. This transformation in scale and culture means the trope of a lone male scientist with unkempt hair soldiering away on a solitary quest is less salient than ever before. Today's big science–engendered era of big data has the potential to accelerate discovery and dislodge established explanations even more rapidly, while also changing the very nature of the questions that scientists can ask and investigate.

We live at a crucial time to understand how science works. I believe a more accurate view of how scientists conduct research and deal with uncertainty will provide greater understanding of the nature of science itself. Studies show that much of the public is ill equipped and unable to craft informed opinions in response to scientific studies, because scientific experts have become increasingly suspect. Complex identity politics, not reasoning, informs belief. Human psychology plays an important role in the acceptance of change. Our attitude toward change is connected at a deep level to our sense of self. In a rapidly transforming world where the frantic pace is triggered by accelerating advances in science and technology, we have a natural tendency to cling to some sense of stability, to believe that such stability gives meaning to our lives. Many recent discussions in the public sphere have rejected scientific findings, designating them "just a theory," as if this were a deficiency. But the beauty of science is that while a theory is always provisional, it represents the best evidence and explanation that we have at any moment. Though prone to revision, science is based on replicable evidence, which privileges scientific over all other possible explanations.

Understanding the power and provisionary nature of scientific thinking is the challenge of our time, and in the pages that follow I offer one cosmologist's view of the complex and contingent side of astronomy. These stories highlight how eminent scientists themselves have struggled repeatedly to accept radical new ideas and how they eventually embraced them. I hope this book will help you to understand (or reaffirm your understanding) that although science as a human endeavor is not entirely objective, it still offers the best prescription for weighing evidence and making sense of the natural world. Shifting and incomplete as it may be, science is self-correcting. It is the best method we have to navigate and make sense of this wondrous universe of ours. For centuries, science has helped us chart our relationship to the natural world. And like any good map, it also points the way forward.

1

EARLY COSMIC MAPS

In the beginning, the only instrument that humans had for observing the cosmos was their eyes. Mythos, not science, governed their interpretations, and they attributed the invisible, mysterious, superhuman forces that guided the planets and the stars to the actions of the gods. When the ancients looked at the heavens, they sought both utility and predictability. And much like we do today, they documented the cosmologies they created. They made maps.

One of the first recorded images of the sky is a hammered copper and gold plate made sometime between 2000 and 1600 BCE, part of the Bronze Age Únětice culture. Discovered in the Saxony-Anhalt region of eastern Germany, it appears to depict the sun or the full moon, a lunar crescent, and stars. To our modern eyes, it also seems to feature the Pleiades—likely, as this star cluster is clearly and prominently visible to the naked eye in the night sky. This metal disk may have been a sort of observational notebook, with new information added over time. One such addition is two golden arcs along its sides, which seem to mark the positions of the sunset at the summer solstice and the winter solstice, thus tabulating the locations of the sun during the longest and shortest days of the year. Another is the arc at the disk's bottom, from which multiple lines emanate and which is variously interpreted as the Milky Way, a rainbow, or a many-oared solar barge, the sun's mythological means of transportation. We know so little about how this object was used. But we

The Nebra Sky Disk (2000–1600 BCE) is an artifact from the Bronze Age Únětice culture, excavated in 1999 in Saxony-Anhalt, Germany. Courtesy Landesamt für Denkmalpflege und Archäologie Sachsen-Anhalt, Juraj Lipták (State Office for Heritage Management and Archaeology Saxony-Anhalt, Juraj Lipták).

can speculate that those who employed it somehow connected what happened on earth to what happened in the heavens.

We also know that the Babylonians, who looked out into the sky some nine hundred years later, were sophisticated recorders of astronomical information. The British archaeologist Austen Henry Layard and his nineteenth-century expedition that aimed to unearth the great biblical cities in Mesopotamia excavated and recovered a rich haul of meticulously tabulated astronomical data. Their find included cop-

The Venus Tablet (seventh century BCE), believed to be part of a longer Babylonian astrology text, *Enuma Anu Enlil*, that connects celestial phenomena to omens. © Trustees of the British Museum.

ies of even more ancient observations that the Mesopotamians had compiled and chronicled. Nestled among the thousands of cuneiform tablets that Layard and his team disinterred in what is now Iraq was a document recording observations of the planet Venus.[1]

Archaeologists believe the Venus Tablet was created during the reign of King Ammisaduqa, and it is just one of several hundred thousand documents that reveal the extent of the Babylonians' interest in recording astronomical data. Translations of the cuneiform show that Babylonians could distinguish between stars that twinkled and planets that shone as steady spots of light. They knew there were five such wandering points, which moved separately from the stars. The English word *planet* reflects this earliest description, originating from the Greek *planētai,* for "wanderer." Relative to other stars, one orb moved from west to east every night. The strangest thing was that about every two years it reversed its motion entirely for some ninety days and then switched back to its eastward journey. The Babylonians recorded this object and its peculiar backpedalling. We now understand this apparent motion of Mars to be a result of the combined movements of that planet and ours—as Earth and Mars pass on their respective paths around the sun, Mars seems to go backward in the sky. The Babylonians were looking for orderliness and had detailed observations of the bizarre motion of the reddish planet, including its unusual backtracking. Comets, which can appear anywhere in the sky and are visible only briefly before vanishing into the darkness, were seen as harbingers of doom, bad omens portending disasters on earth. From their detailed chronicling of the movements of orbs in the night sky, it is clear that many ancient civilizations noted the regularity of the heavens and strove to predict future positions. Successfully doing so probably helped them to come to terms with nature. The maps of the ancients drew connections between the celestial and the terrestrial.[2]

Today we use astronomical data to support or overturn astrophysical concepts and models, but in ancient times human understanding of the heavens had a more intimate connection to quotidian events. Registering current celestial events was in the service of predicting future ones, but the ancients were not seeking to explain patterns or to arrive at their causes. Their goal was to record movements and to develop descriptions that would enable accurate future prediction. This is the root of astronomy—observation. Seeing and recording how objects move in the sky eventually gave birth to a science, even if the original explanation for these objects' movements was anything but scientific. This early tradition that centered on taking data from the night sky was crucial. It gave society an instinct—to connect our place on our planet to our location in the cosmos.

Despite the Babylonians' inability to scientifically understand the motion of the wandering orbs, their observational data had practical and religious purposes—patterns in the sky, for example, were of great importance to agricultural cycles on the ground. Consider this observation from the Venus Tablet: On the fifteenth day of the month, Venus disappeared from the heavens and remained unseen for three days. Then on the eighteenth day of the eleventh month it reappeared in the eastern sky. New springs began to flow, the god Adad sent rain, and the god Ea sent his floods.[3] The retrograde motion of Venus meant downpours on earth. In Hindu mythology, Indra, the supreme deity and god of storms, is variously referred to as the Lord of Lightning, the Storm Gatherer, and the Bestower of Rain. He is eternally engaged in fighting demons from the underworld and battling evil on behalf of the forces of good. He is the demiurge—an artisan or worker figure who was believed to fashion and maintain the physical universe, responsible solely for the material world, not the creator—who pushed up the sky and released dawn, therefore requiring appeasement to keep up the regularity of night and day.

Because the data itself was not used to reveal physical causes at that time, the ancients, lacking advanced technology and theory, invented astrology. Ancient Indian astrological tradition, for example, partitioned the night sky into houses of the zodiac, replete with elaborate mythological stories that accounted for their shapes. Each planet had a ruling lord and an associated temperament. Mars, for instance, was warriorlike and made its natives (those born in its portion of the natal chart) aggressive, argumentative, lovers of weapons, and bestowed with technical and mechanical abilities.

The shift to a world view rooted in logic, data, and evidence had to wait for the ancient Greeks. The origin story that held sway when they appeared on the scene was one in which the world rested on the back of a turtle, which was supported below by yet another turtle . . . with turtles all the way down. This image (sometimes with minor variations) was the prevalent belief right up to the sixth century BCE. But compared to the established cities and kingdoms of antiquity, such as Jerusalem and Babylonia, there was something radical, novel, and dynamic about the emerging Greek world. Unlike ancient kingdoms, it consisted of several politically independent city-states, autonomous and fragmented. Marked by openness to question and debate, this burgeoning culture remade the pantheon that ruled the heavens. Gods were refashioned and more power and agency transferred from the divine to the human. In fact, the divine even reflected human flaws, narrowing the schism between godly perfection and human imperfection.

Into this setting, in 610 BCE on the Ionian coast in the town of Miletus, in present-day Turkey, Anaximander was born. He is attributed with seeing the earth as a cylinder floating in space surrounded by the heavens, not held aloft by any creature. He is acknowledged as the first person to deduce that the earth was freely suspended. This was a profound change in world view, an incredible leap that was emblematic of his entire take on the cosmos.

Supremely radical though this was, it wasn't just what Anaximander thought about the relationship between the earth and the heavens that was so transformative, but also the intellectual process by which he arrived at his ideas. Although his teacher Thales is credited with abandoning mythological explanations, Anaximander is said to have set in motion the rethinking of our world, igniting a search for knowledge grounded in questioning and challenging what appeared to be fixed and certain. This kind of inquiry is a prerequisite and defining element of any form of critical thinking—but especially our current scientific approach. Anaximander's attempt to explain nature and elucidate the origin of humans and the world with one encompassing account was one of the most imaginative and earliest, if not the first. If there is a moment in history that we can pick out as a turning point, it was then—when Thales and Anaximander, both residents of Miletus, were formulating a radical new world view. Anaximander did not passively accept the status quo; he sought knowledge and realized that it was continually evolving. His understanding was neither absolute nor static. It required questioning, scrutiny, and continual reformulation.[4]

One of the roots of the critical thinking that is the linchpin of all astronomy is the desire to question, driven by curiosity. Another is humanity's continued desire to know and to represent what is known as a map. We cannot underestimate the importance of this literal and practical tie between the heavens and the earth, which developed with geodesy—the science of global positioning. One tool that eventually proved crucial for geodesy was the magnetic compass invented by the Chinese in about 200 BCE. Made of lodestone, a naturally occurring magnetic material found in the ore magnetite, these compasses align with the earth's magnetic field. But at that time, lodestones were used only for purposes like feng shui, harmonizing oneself with the surrounding environment. It was not until around 1040 CE that the Chinese themselves used compasses for terrestrial navigation or military purposes, and it took another hundred years before their use spread to maritime navigation. How the knowledge of magnetism disseminated from China to the West is still a matter of debate among historians, but there is ample evidence that it was the Chinese who invented the compass. So the rest of the world had to wait for the compass, and the major impetus for mapping's advancement in ancient times required more than the terrestrial—the heavens had to be mapped. Stars at night helped the ancients to navigate the oceans, and our sun allowed them to measure the size of the earth.[5]

One of the early milestones in mapping was the estimate in 240 BCE of the circumference of the earth by the Greek astronomer Eratosthenes, who noted that in the town of Syene (modern-day Aswan) on the longest day each year (the summer solstice) there was no shadow cast at noon. He was aware that the sun was not directly overhead on the same day in his hometown of Alexandria along the Nile in northern Egypt, so he estimated the discrepancy in its position by calculating the angle produced by the shadow of a tall tower in Alexandria. Using geometry and knowledge of the distance between the two cities, he arrived at a figure for the circumference of the earth that was just about 16 percent off the modern measurement of about forty thousand kilometers (twenty-five thousand miles).

Mathematics allowed a radical new approach to thinking about the cosmos—a move away from mythos to logos, to a physical and geometrical conception of motions that allowed the characterization of regularities. Hipparchus of Nicaea (190–120 BCE) is considered one of the greatest ancient astronomers. Many credit him with the invention of trigonometry and with producing the first effective models for the motion of the sun and the moon. He likely had access to Babylonian records of eclipses and positions of the planets. Building on work they and the Mesopotamians had done, Hipparchus assembled an up-to-date star catalog and developed what are believed to be the first quantitative, geometrical and mathematical descriptions of astronomical data. In the second century CE, the Greco-Egyptian astronomer, mathematician, cartographer, and astrologer Claudius Ptolemy took the next major step toward un-

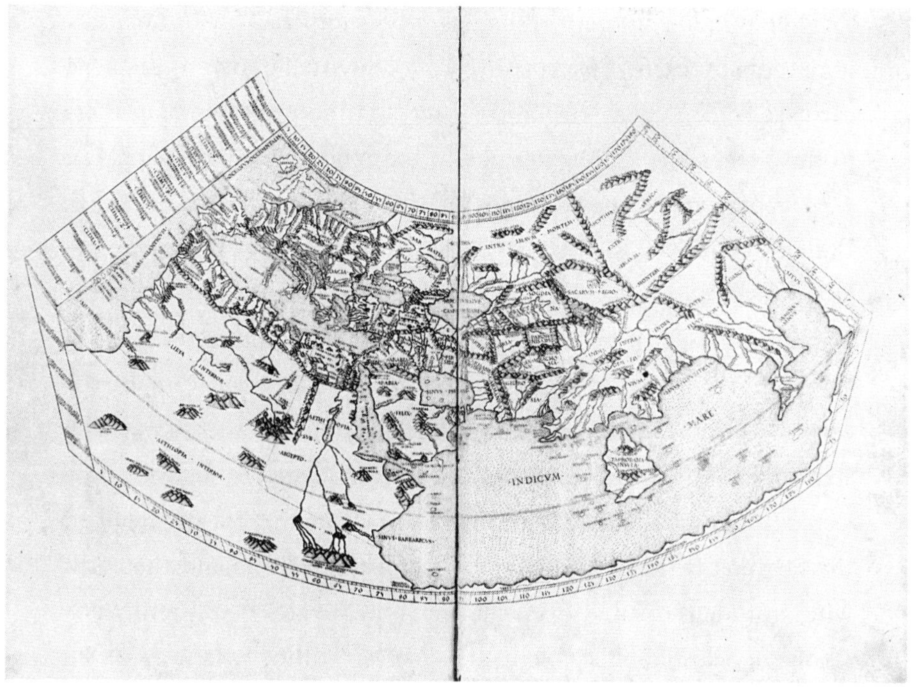

A copy of Ptolemy's world map (1478). Ptolemy knew that the earth was a sphere, and his projection measures latitude from the equator, a convention still in use today. Courtesy of the Library of Congress.

derstanding the motion of heavenly bodies. Inheriting Hipparchus's three-hundred-year-old data, he collated all the astronomical tables and geometrical models of the Greeks into a comprehensive treatise, the *Almagest*. But he did more than just compile. He constructed a new model of the heavens, which was consistent with all the available data.[6]

Ptolemy's physical model consisted of nested spheres, and his comprehensive tables allowed the calculation of the future positions of planets. He used four observations spread out in time for each planet to obtain maximum leverage on the estimate of their periods. The oldest observation that he used, from 700 BCE, most likely came from Hipparchus's compilation of Babylonian records. Given that the primary interest in planetary positions still lay in forecasting events on earth, unsurprisingly Ptolemy was drawn to map the earth as well. While his *Almagest* records the positions of planets in the heavens as well as periods of the moon, its counterpart, the *Geographica,* maps the locations of cities and landmarks on earth. Ptolemy conceived the two in tandem: just as he had ordered the celestial realm in enclosed spheres, he assigned locations to all known places on the earth on a grid. Since the planets and the sun move along the ecliptic, Ptolemy used ecliptic coordinates—an earth-centered grid as seen from outside the celestial sphere—to map the star catalog. From this time on, both the earth and the sky were projected with coordinates on the surface of a sphere. Ptolemy mapped the heavens with the stable reference of the ecliptic and the earth with latitude measured from the equator. The ability to predict positions of celestial objects lent the *Almagest* authority that prevailed throughout the Middle Ages.

The Greeks had also devised the mathematics to study arcs, portions of a circle and the angles subtended by chords that connect the center of the circle to the circumference. But the development of mathematics of course benefited from those beyond ancient Greece's borders. The Indians expanded Hellenistic mathematics, and in particular the mathematician Aryabhata, working in the fifth century CE, is credited with describing trigonometric functions via infinite series, enabling him to develop extensive tables of values for the sines and cosines of angles. To map the heavens on a celestial globe and the earth on a terrestrial globe, two-dimensional Euclidean geometry needed to be extended to curved surfaces. The Arabs and Indians developed spherical trigonometry from the seventh to the eleventh century. The extension of geometry to describe the relationships between the sides and the angles of triangles on the surface of a sphere was critical to astronomy, to map the locations of stars on a globe; and to geodesy, to understand the impact of the curvature of the earth on navigation as distant locations on earth were now becoming accessible.

Thriving sea trade routes brought Persian and Arab mathematicians into constant contact with the principles of Indian mathematics, which they translated and widely disseminated throughout the medieval Islamic world. The mathematician al-Jayyani of al-Andalus wrote what is believed to be the first comprehensive treatise on spherical trigonometry, *Kitab majhulat qisiyy al-kura* (The book of unknown arcs of a sphere). Applying Ptolemy's theorem that gives the longitudinal difference between two places on earth in terms of their difference in latitude and the great circle distance between them, the mathematician Rayhan al-Biruni used caravan routes to derive the differences in longitude between Baghdad and other cities in the eleventh century.[7]

Astronomy requires synthesizing observations with a theoretical and mathematical framework, and the reasoned analysis of cause and effect. Although Ptolemy's model could map the motions of the planets and chart the brightest stars recorded by the Babylonians, he was not in search of an explanation for our subject of interest: the cause of the motion of the planets.

Technology, again, was key, and it had been advancing. The compass, an invention of 200 BCE, appeared in the Western world some fourteen hundred years later. In *De naturis rerum* (On the nature of things), Alexander Neckam mentions the magnetic compass and its use in navigation. This is about forty years prior to its reference in a Persian book of tales dating to 1232, the *Kitab kanz al-tujjar fi ma'rifat al-ahjar* (The book of the merchants' treasure), written by Baylak al-Kibjaki in Cairo.[8]

These concrete representations of the refinements in mathematics and cartographic instruments finally produced a radically new

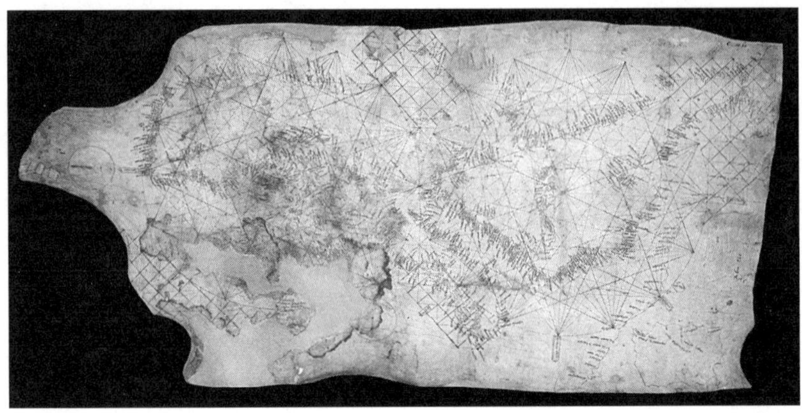

The *Carta Pisana* (1296) is the oldest surviving nautical chart, covering an area from the present-day Netherlands to present-day Morocco. This portolan map provides a detailed survey of the coasts and port cities that it displays and was drawn to scale. Courtesy of the Bibliothèque Nationale de France.

kind of map—one drawn to scale. Portolans combined directions derived from the compass and distances estimated by sailors at sea. The development of these charts spurred the so-called Age of Discovery, which was the start of the Age of Accuracy for Astronomy. The resulting quest for power and plunder among seafaring Europeans encouraged all manner of scientific and instrumental innovation. As their name, deriving eventually from the Latin for "port," implies, the focus of portolans were the details of coastlines and the routes that they mapped out—lines connecting known coastal cities were drawn that enabled the calculation of the distance as well as the time that the journey would take. The oldest surviving portolan is the *Carta Pisana,* which dates to 1296.

While portolans embodied the desire for greater scientific accuracy, by using stars to map the earth, celestial maps not only grew more accurate at this time but also began to convey more sharply and compellingly the shifting explanations for cosmic phenomena. This change in explanatory methods and power, which reflected important conceptual transformations, is nowhere more obvious than in maps of the sky. Consider for example the map of the cosmos that appears in *Le breviari d'amor* (The abstract of love), an illuminated manuscript attributed to Matfre Ermengau of Bezier and published between 1375 and 1400.[9]

This depiction incorporates the Aristotelian and Ptolemaic views of the universe; the realm of the fixed, unchanging, and perfect stars is clearly demarcated in the outer rim. All imperfections are restricted to the earthly sphere, within which lie the mutable elements—fire, water, earth, and air. Everything else is assumed to be pure and perfect. Note how this image adopts a depiction of a divine agent coupled with a mechanistic understanding: the daily motions of the sun and moon are shown as caused by laboring angels that fuel the earth's rotation. So here we have a well-ordered Ptolemaic cosmos that is nevertheless powered by angels, depicted as turning a crank, clearly used as a metaphorical device. This map reveals the persistence of mythos, or spiritual elements, coexisting with a mathematical representation. Here angels occupy the explanatory vacuum that Isaac Newton's laws of gravitation later filled. Newton, of course, saw gravity not as a property of matter but as a manifestation of the divine. He believed in spiritual forces as the drivers of planetary motion.

Representations of the cosmos grew commensurately more elaborate with improved understanding. Likewise, changes in views of the cosmos also got depicted cartographically. Some of the extremely gorgeous renditions of the medieval cosmos can be found in the *Catalan Atlas,* published in 1375. This is one of the most significant compilations from the Middle Ages that depict conceptions of earth and sky. This atlas is attributed to the Jewish astronomer and cartographer Abraham Cresques. The earth in this image is surrounded by rings, which represent the four essential elements and seven spheres, which denote the orbits of the then-known planets. Beyond these lie

Illuminated map from Matfre Ermengau of Bezier's *Le breviari d'amor* (The abstract of love; 1375–1400). The image depicts the Aristotelian-Ptolemaic cosmos, wherein everything contained within the sphere of the moon was mutable and corruptible, while all celestial phenomena beyond the lunar orbit were pure, unchanging, and perfect. Here eternal angels turn the crank to rotate the sublunary sphere perpetually. © The British Library Board.

the moon, the sun, and the fixed stars. This map signals the transition from the age of the angels to the age of instruments. Angels are no longer invoked to power the cosmos; instead we have the growing influence of scientific instruments—notably the astrolabe held by a sagelike figure who appears prominently at the center of the map.

Abraham Cresques's illuminated map of the cosmos from the *Catalan Atlas* (1375). This medieval depiction shows our planet surrounded by rings signifying the four elements—earth, air, water, and fire—beyond which lie seven spheres, depicting the orbits of the planets, followed by the moon, the sun, the sphere of the fixed stars, and the zodiac. Here Cresques has replaced God with a sage holding an astrolabe, a likely reflection of his personal view of the cosmos. Courtesy of the Bibliothèque Nationale de France.

Although the astrolabe as a device for measuring positions is said to have been invented by the ancient Greeks and is often attributed to Ptolemy, it was refined in the medieval Islamic world. Medieval Islamic scholars and their knowledge of trigonometry led to the inclusion of angular scales on the instrument. The astrolabe was used to determine the positions of the sun, the moon, and the stars as well as local time at a given latitude using the tabulation of latitudes of many prominent cities contained in the device as individual detachable plates. In the Islamic world, the spherical astrolabe was also used to determine the direction toward Mecca and to find the times for daily prayers for the devout. The first Western astrolabe made of metal was fabricated in tenth-century Spain, so it is not surprising that the instrument makes an appearance in the *Catalan Atlas*. In Cresques's map, time is now a mathematical concept that can be measured to eternity. The power of mathematical calculation is front and center. In prior depictions of the cosmos, bearded wise men representing gods always appeared in the image, controlling the scene. On the eve of the Renaissance, angels and cherubs are absent, replaced by allegorical human figures that represent the four seasons.

The Renaissance astronomer Nicolaus Copernicus took a radical next step in 1514 in a manuscript that was about twenty pages long and, in a sense, a preview of coming attractions. Later titled *Commentariolus* and circulated only among his friends, this work revamped the prevailing, Ptolemaic view of the cosmos. Copernicus proposed a reordering of the heavens, creating a new reference system that now placed the sun at the center rather than the earth.

There is no doubt that the Copernican system was a disruption of all prior notions of the heavens—it implied not only that the earth revolved around the sun but also, because the arrangement of stars remained fixed despite the earth's movement in its proposed orbit (an absence of what is called parallax), that the stars were far, far away. The celestial frontier had been pushed out. Fearful of re-

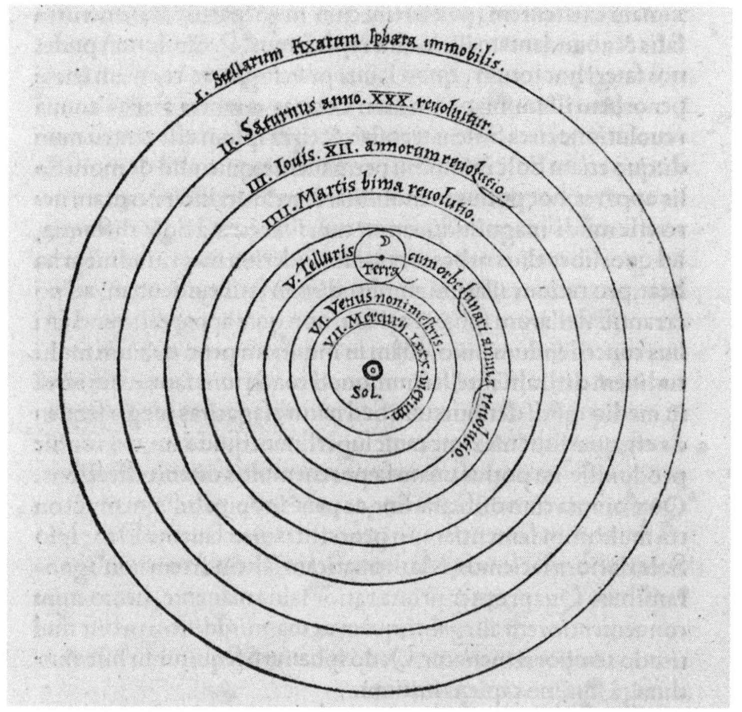

Nicolaus Copernicus's heliocentric model, from *De revolutionibus orbium coelestium* (*On the Revolutions of the Heavenly Spheres*; 1543). Copernicus boldly moved the earth from the center of the Ptolemaic cosmos, replacing it with "Sol"—the sun in this simple diagram. Around the diagram, in his elegant hand, Copernicus explains his model. Courtesy of the Library of Congress.

jection, Copernicus at first hesitated to publish his full treatise on the subject, *De revolutionibus orbium coelestium* (*On the Revolutions of the Heavenly Spheres*), until 1543. Eventually, a bishop encouraged its publication, and Copernicus dedicated it to the pope. It was only more than seventy years later, in 1616, that the work was banned, until it was "corrected" by the Catholic inquisition. A list of "corrections" was issued, with a handful of passages (in about ten places) deleted, including ones that presented the motion of the earth as a fact rather than a hypothesis. These amendments were made to present heliocentrism as just a convenient way of describing the planets' motion—a frame of reference—not reality. As we will see in many of the following chapters, similar dodges have often been needed as part of the course in making radical ideas more palatable.

The eminent historian of astronomy Owen Gingerich tracked down almost every existing copy of Copernicus's book and, with a bit of clever detective work, estimated the surviving fraction. He concluded that four to five hundred copies were likely printed in the first edition and a further five hundred or so in the second edition, issued in 1566. Gingerich recounts his quest to find these copies in his ironically titled *The Book Nobody Read* and notes that half of those in Italy were amended, while very few were elsewhere in continental Europe.[10]

Though the influential Cardinal Bellarmine eventually spearheaded the charge against heliocentrism, discomfort with the reshaped cosmos was not exclusively Catholic. In fact, Martin Luther objected to heliocentrism. Of course, church doctrine at the time held that the earth and not the sun was fixed at the center of the cosmos. Copernicus's map, though it explained the general motion of the other wandering orbs, was unable to predict the erratic motion of Mars or Venus any more accurately than the prevailing model. And for deniers, the lack of parallax—that perspective shift that should result from the earth's change in location—could mean not that the stars were far away but, more simply, that the earth was not moving at all. Copernicus's rearrangement of the cosmos was a brilliant stroke of imagination. It was definitely not driven by data. Technology was partly to blame. Observations were so inaccurate that even broad-brush forecasts were acceptable, and there had been no real improvements in measurements since Ptolemy.[11] A new cosmic view required more accurate data to support it. Incidentally, as was common for his time, Copernicus, was interchangeably referred to as an astrologer or an astronomer—although he never cast horoscopes.

The new primacy of empirical data marked an important turn in the history of science as well as in the history of ideas in cosmology, and it set a new standard for epistemology. It also marked the move from the ethereal to the material in knowledge acquisition. Astronomy was at the forefront of this empiricist revolution. Observers were able to make repeated observations over time and determine underlying patterns, and this also set the stage for the development of an intellectual scientific community. The invention of the printing press made the rapid dissemination of information possible and offered a new means of communicating ideas, opening up a dialogue among practitioners. Astronomers wrote books that were printed and then circulated among other active astronomers.[12]

Various printed maps and other depictions of the cosmos from the sixteenth and seventeenth centuries provide evidence of the resulting conceptual tussle between competing celestial models. It took Tycho Brahe, the sixteenth-century Danish astronomer, to come along and revolutionize the field. With ample resources to continually build and refine astronomical instruments, he was obsessed with improving the accuracy of observations. He was very organized, and to him observations were essential. He launched specific campaigns and efficiently collected data when the planets were in interesting geometric configurations, for example in opposition. Brahe continued to amass observations to support or refute older models. He was the last of the great naked-eye astronomers. He observed comets in detail, and this led him to dismantle the then popular Aristotelian view of a perfect, fixed, and unchanging universe beyond the orbit of the moon. Even as he challenged the old paradigm, he was uncomfortable with the earth-sun swap that Copernicus proposed. He crafted an alternative system, wherein all the planets—except for the earth—orbited the sun, which in turn orbited the earth with all its planets in tow. Andreas Cellarius's *Harmonia macrocosmica* (Cosmic

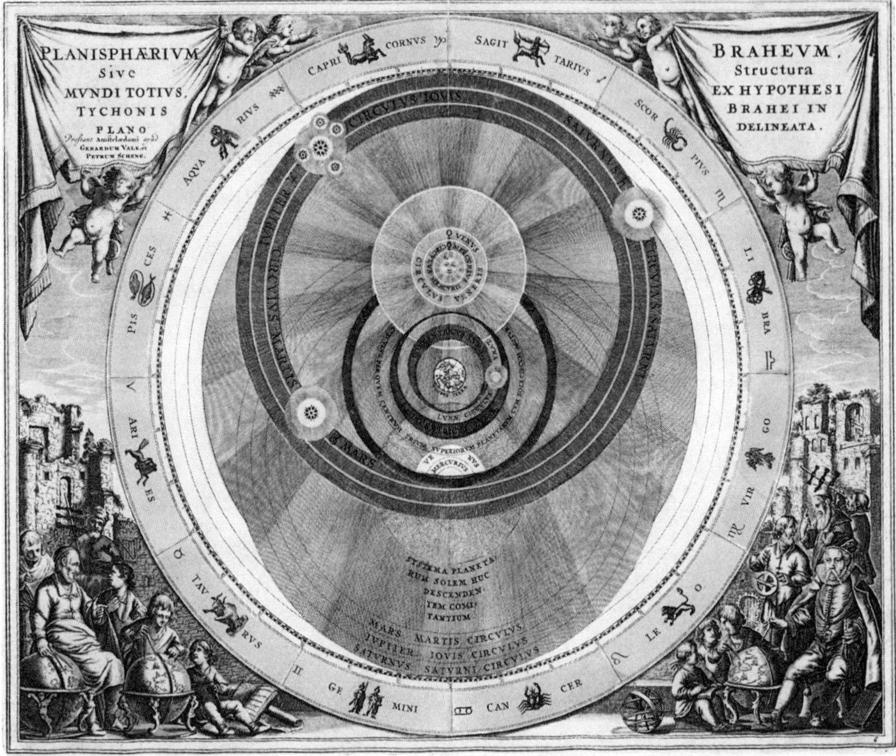

Map of Tycho Brahe's modified geocentric model, from Andreas Cellarius's *Harmonia macrocosmica* (1708). In this model all planets other than the earth orbit the sun, while the sun orbits the earth. Courtesy of the Stephen S. Clark Library, University of Michigan Library.

harmony) depicts this view. Such a middle-ground model is a typical dodge when a radical idea challenges leading minds. Often what causes the final shift is not a single attributable event or an identifiable tipping point but rather a slow and steady accumulation of solid supporting evidence that finally changes minds.

The debates that ensued between the proponents of Copernicus's and Brahe's models and their respective views of the cosmos were the subject of many artistic depictions. Maps reflect the battle between these conceptions. They became sites for the dissemination of new ideas as well as instruments of intellectual persuasion.

Consider for example the Italian astronomer and Jesuit priest Giovanni Battista Riccioli's adaptation of Brahe's model, discussed in his treatise *Almagestum novum* (New almagest). An illustration of Urania, the divine muse of astronomy, serves as a frontispiece to the book. In the image, she literally weighs the Copernican system, on the left, against Riccioli's adaptation of Brahe's model, on the right. The scales in Riccioli's book (of course) fall in favor of his own theory, in which Mercury, Venus, and Mars orbit the Sun, which orbits Earth, like Jupiter and Saturn, which remain in their Ptolemaic geocentric orbits. To the left is the many-eyed Argus, holding a telescope and pointing at the plethora of new celestial objects that it has brought into view. We also see the sagelike Ptolemy, reduced to a spectator with his discarded geocentric model on the ground.

Beyond the map, theological affiliations and political allegiances also influenced Urania's balance. Aside from the intellectual objection of not finding evidence for parallax, Brahe's anti-Copernican view had the political advantage of being in consonance with Catholic dogma, which ordained the earth immovable. This emerged from a literal reading of the Bible, a new practice that arose in response to the challenges posed by the Reformation. A large number of seventeenth-century astronomers who were uncomfortable with Copernicus's picture bought into Brahe's view. But Brahe's competition soon came from close quarters—his colleague and scientific collaborator Johannes Kepler.[13]

Again, the red planet played a prominent role, although the key problem pertained to the placement and shape of Earth's orbit. Kepler solved this bigger problem by using Brahe's comprehensive data on Mars. Since it is the planet closest to Earth, the inaccuracy resulting from the uncertain position of Earth's orbit appears most glaringly in the calculation of Mars's position. In his 1595 book *Mysterium cosmographicum* (The cosmographic mystery), Kepler published a defense of the heliocentric model postulating that the planetary

Frontispiece of Giovanni Battista Riccioli's *Almagestum novum* (New almagest; 1651). The muse of astronomy, Urania, is weighing the Copernican model against Riccioli's, wherein Mercury, Venus, and Mars orbit the Sun, which in turn orbits Earth, while Jupiter and Saturn remain in their Ptolemaic geocentric orbits. Courtesy of the Library of Congress.

spheres, centered around the sun, could be inscribed with combinations of Platonic solids. The structure he proposed was a system that resembled Russian nesting dolls. He also generated the next major and most radical shift yet—the search for laws, perennial truths that could be derived to describe and account for the motions of the celestial bodies. Kepler strived to develop celestial physics—he attempted to derive a physical theory that would provide an explanation for and describe the causes of the motion of the planets. The geometric and complicated world view at which he arrived seems at odds with his penchant for pure mathematics, but the same deductive powers required to visualize this complex scheme led him to postulate the three laws of planetary motion. Kepler failed to arrive at the concept of inertia and instead invoked the rotation of the sun as the continual and dynamic source of energy that holds the solar system aloft. His three laws predict (1) that the orbits of planets within the solar system are ellipses; (2) that in an elliptical planetary orbit with the sun at one of the foci, a line segment joining a planet and the sun sweeps out equal areas during equal intervals of time; and (3) that there is a direct relationship between the orbital period of a planet and the size of its elliptical orbit (specifically, the period squared is proportional to the semi-major axis cubed). There was now an explanation for the motions of the planets—still devoid of physical causes—and divine order was still an inseparable part of the scheme.

Up to this point, the position of the earth with respect to the sun—and hence the earth's orbit—was incorrect. The eccentricity of its orbit was off by a factor of two even for Copernicus. More accurate data from Brahe provided the adjustment to the earth's orbit, and this was crucial to Kepler's inference of motions in ellipses.

Following the correct placement of earth's orbit and the formulation of Kepler's laws came the solution to the puzzle of Mars. Earth

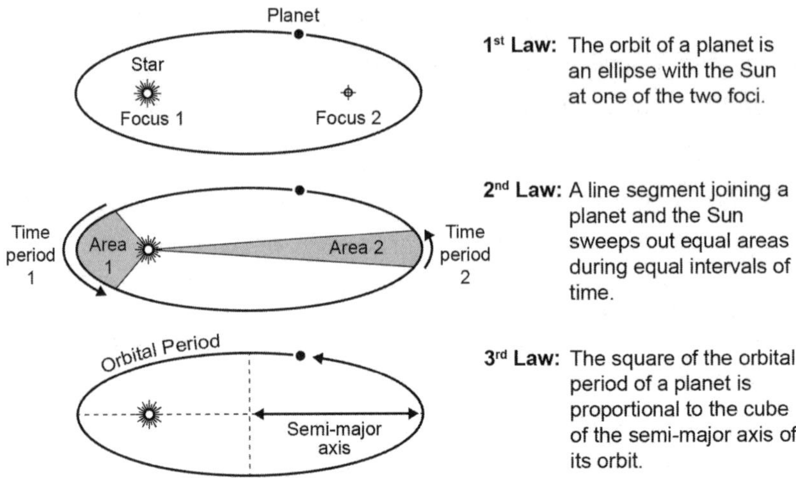
Planet
Star
Focus 1
Focus 2
1st Law: The orbit of a planet is an ellipse with the Sun at one of the two foci.
Time period 1
Area 1
Area 2
Time period 2
2nd Law: A line segment joining a planet and the Sun sweeps out equal areas during equal intervals of time.
Orbital Period
Semi-major axis
3rd Law: The square of the orbital period of a planet is proportional to the cube of the semi-major axis of its orbit.

Using Brahe's extensive observational data and seeking a physical explanation for the motions of planets, Kepler formulated three laws that govern planetary motion.

and Venus revolved around the sun in orbits that deviated only very slightly, almost imperceptibly, from a perfect circle, which agreed with the Ptolemaic picture. Mars, on the other hand, has a much higher eccentricity, which a circular orbit cannot accommodate.

Kepler was a convinced Copernican and never embraced the hybrid model that Brahe proposed. But even he had no real explanation for why the planets moved, beyond Ptolemy's idea that a "prime mover" spurred the celestial spheres. Kepler, though, was the first to search for causation, as we now understand it in modern scientific terms. He insisted on the notion of having a physical theory and attempted to develop the principles of celestial physics. Aside from the rotation of the sun, he considered magnetism as the potential organizing force of planetary motions. In the classical view, even up to Copernicus, there was never an attempt to look for a physical reason why the planets moved as they did. Despite this pioneering effort, Kepler came up short, as he did not understand the role of inertia. The cause was seen as a philosophical rather than astronom-

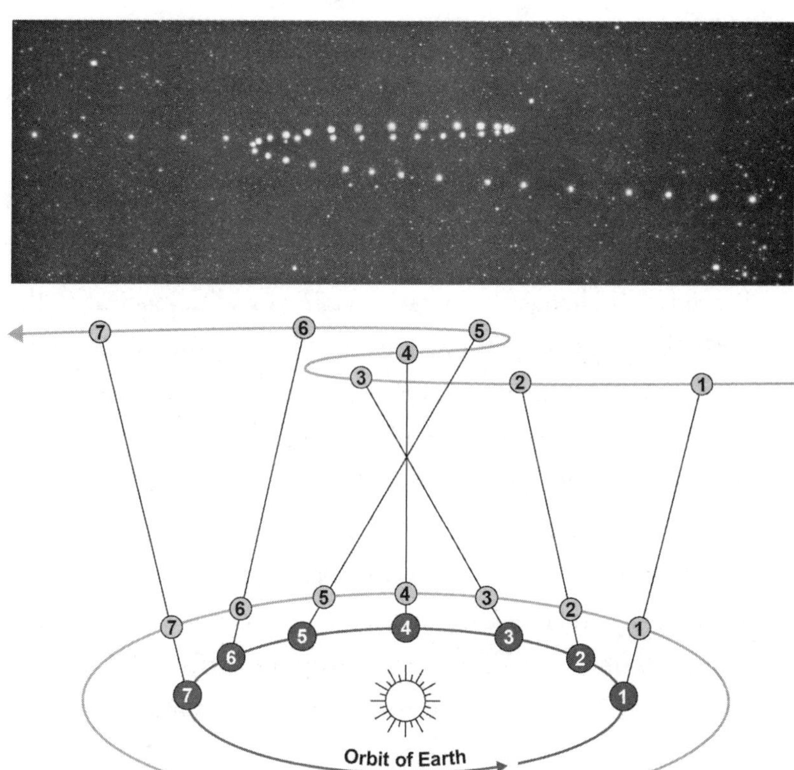

Viewed from the vantage point of Earth, in its orbit around the Sun, Mars appears to travel backward periodically in the sky, only to resume its forward motion. Copernicus's heliocentric arrangement of the heavens coupled with Kepler's ellipses was finally able to account for this erratic movement.

ical issue. Of course, astronomy was part of natural philosophy. In many ways astronomy proved to be the intellectual discipline that catalyzed the cleavage of natural philosophy to form what we now refer to as modern science.

Retracing the genesis of a new idea is a challenge. As we see in the evolution of models that I have just described, maps show us the state of knowledge at a particular point in time and serve as powerful markers of the dynamic by which new ideas are introduced, circulated, debated, and contested. They marry observation, technology, and understanding.

While the ancients had only their eyes to rely on, modern astronomers have telescopes on the ground and in space to extend their vision and observe the near and distant universe. Celestial maps bear the imprint of this transformation, charting how the human view of the heavens has changed from the imagined and fantastical to the reasoned. Although Kepler provided a convincing model, it took a new scientific instrument and a new idea to settle the issue of the planetary motions once and for all. Marketed as a spyglass in Amsterdam in 1608, on being repurposed, the telescope brought distant objects in the night sky into view. Galileo Galilei is credited with the invention of the astronomical telescope, an improvement on this simple spyglass, which he used to discover the moons of Jupiter, sunspots, and the phases of Venus, and to map the surface of the moon. Galileo also helped advance the notion of celestial physics. The next big step had to wait for the English physicist Isaac Newton and his 1687 publication, *Philosophiae naturalis principia mathematica,* often simply called the *Principia,* which outlines the universal law of gravity. This treatise would have been inconceivable without the discovery of Kepler's three laws. Newton made the boldest leap thus far, uniting the terrestrial and the celestial with a universal law of gravitation. He wiped out the distinction between the heavens and the earth and showed that the same rules work on both. The methodological program of what we refer to as science began to emerge at this time in the 1600s.

The radical transformation in world view—heralded by Copernicus and supported by the telescopic observations of Kepler, Galileo, and many others—revived another ancient speculation about the structure of the larger cosmos. This development led to the resurgence of interest in whether there were other worlds beyond our

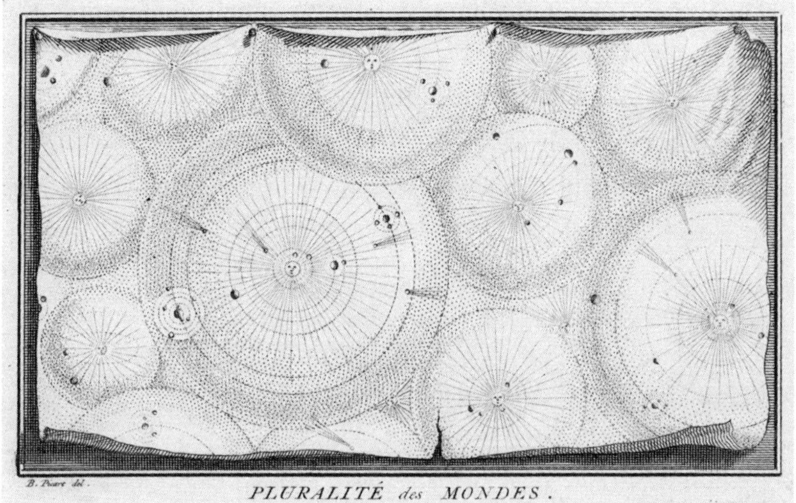

The French engraver Bernard Picart's 1673 depiction of the plurality of worlds. Courtesy of the Rijksmuseum, Amsterdam.

solar system. A late seventeenth-century print by the French master engraver Bernard Picart reveals his idea of the notion of the plurality of worlds that could exist throughout the universe—a multiplicity of other stars besides our sun that could harbor their very own planetary systems, mirroring our solar system. With the solar system settled, astronomers leaped beyond it to reconceive and map out what might lay farther afield.

Along with the rest of today's astronomers, I have inherited this ancient preoccupation and legacy of map making. Although we may model using our computers instead of astrolabes, we remain explorers of the cosmos. The frontier is no longer the edge of the world, explored aboard a caravel, but instead the edge of our universe, seen through humanity's most powerful telescopes. We draw and redraw our cosmological maps with the help of these increasingly sophisticated instruments. We see new frontiers that stretch well beyond our imagination—to the farthest reaches of space, and back to the hiss of the infant universe soon after the big bang's moment of creation. We continue the tradition that began as we moved to a world governed by logos and finally by the scientific method. We will see that progress in the pages that follow—new observations and new theories contest and refine the radical conceptions of place that research in cosmology has wrought.

2

THE GROWING BORDER

The Universe Expands

On a cold February morning in 1848, Edgar Allan Poe gave a lecture titled "On the Cosmography of the Universe." The location was the stately Society Library in New York. Only sixty people attended, and they left utterly disappointed and baffled. Still, this talk and the work that preceded it would serve as the basis for the prose poem "Eureka," in which Poe presents his personal conception of the origin of the universe. Some read "Eureka" as prophetic—anticipating new scientific discoveries—others as romantic, deeply personal, or even intentionally satirical. In the opening section, Poe declares, "I design to speak of the Physical, Metaphysical and Mathematical—of the Material and Spiritual Universe: of its Essence, its Origin, its Creation, its Present Condition and its Destiny." He goes on to describe the universe as restless and evolving. This was in stark contrast to the scientific community's existing view of a static universe. Lacking proof, Poe's poem attempts to convince through suggestion. By 1848, however, it was impossible to persuade anyone of the validity of a new scientific idea without presenting empirical evidence. A scientific explanation needed support from data. Of course, Poe had not conducted any scientific research. But he was right.

It took astronomers more than eighty years to support Poe's claim. In 1929, Edwin Powell Hubble, using the state-of-the-art hundred-inch telescope on Mount Wilson in Southern California, uncovered an incredible correlation: the farther a galaxy was from us, the faster it appeared to be hurtling away. His observation made sense only if the universe was indeed expanding. "Eureka" was confirmed! This discovery caused a radical shift in our conception of the cosmos, as dramatic as the one that Nicolaus Copernicus's 1543 heliocentric model produced. The idea of the expanding universe garnered support and marked the emergence of a fundamentally new picture of the cosmos. Our cosmic map got a twentieth-century makeover.

If this story starts with Edwin Hubble, then Albert Einstein plays an important foil for our protagonist. At the same time that Hubble, the astronomer, was busy dislodging the universe with his observations, Einstein, the famed theorist, clutched on to the idea of a fixed universe. The tussle here was not between Einstein and Hubble as individuals, or even between theory and observation, but between belief and evidence. On another February day, this time in 1931, at a seminar at the Mount Wilson Observatory—fittingly, as it was the site where Hubble took his data—Einstein finally admitted that he had been wrong, an assertion that shocked all the attendees assembled in the room, including Hubble. A reporter from the Associated Press wrote that "a gasp of astonishment swept through the library."[1] That gasp is emblematic of the human dimension of scientific discovery.

But now that I've given away the punch line, let's go back to Hubble's beginnings. On May 6, 1906, the handsome sixteen-year-old senior from Wheaton High School in Chicago broke the Illinois state record for the high jump. The *Chicago Tribune* reported that the young Edwin Powell Hubble had cleared a six-foot-one-inch bar, perhaps more legend than reality. As Alan Lightman recounts, later that year Hubble allegedly won medals for everything from the pole vault to the shot put to the discus and even the hammer throw. When he won a scholarship to attend the University of Chicago, he seemed well on his way to life as a professional athlete. Hubble was well built, tall at six feet two inches, and incredibly ambitious. In addition to having an attractive physique, he was mentally agile, academically gifted, and, from his sister Lucy's account, arrogant from an early age. Despite being prone to exaggerate his abilities, Hubble was intellectually curious and widely read. He developed an early interest in astronomy when, at the age of eight, he received a telescope from his grandfather William James. His first glimpse of the heavens seems to have left an indelible impression. After a brilliant undergraduate academic career at the University of Chicago, he won a Rhodes scholarship to study at Oxford. Winning the Rhodes had been an important goal for him, and living in England made him a lifelong Anglophile. To please his father, Hubble studied jurisprudence at the Queen's College, giving up his dream of pursuing either astronomy or mathematics while in the United Kingdom. His contemporaries, other scholars awarded the Rhodes in the same year, were the newsman Elmer Davis, who went on to head the U.S. Office of War Information during World War II, and the mathematician and electronics pioneer Ralph Hartley. In his time at Oxford, Hubble dandified, procuring a chipper, upper-class English accent, and developed affectations that were in keeping with the aristocratic set with whom he mingled. He maintained these scrupulously for the rest of his life, including smoking a pipe, even while observing at Mount Wilson in his later years.[2]

He returned to the United States in 1913, rejoining his family and supposedly opening a law office in Louisville, Kentucky, but it turned out that he had merely suspended his dreams of the skies. Although he wrote back to his friends in England that he was handling legal cases, Hubble was actually teaching high school physics, mathematics, and Spanish in New Albany, on the other side of the

Ohio River from Louisville.[3] His father had died earlier that year, so he had returned to help his widowed mother and siblings. Hubble was devastated by this loss. However, he also felt released from the oppressive expectations of his stern father. He quit working within a year of returning from England and went back to the University of Chicago, where he enrolled as a graduate student in astronomy.

Before Hubble's discovery, every civilization and mythology in every part of the world believed unstintingly in a universe that held steady—and was unchanging. In creation myths across millennia, cultures had grappled with the ever-changing natural phenomena on earth—rain, thunder, lightning, floods, and drought—by invoking a fixed heaven, a static cosmos. Our ability to see the unchanging stars in the night sky of course supported this belief.

In *On the Heavens*, Aristotle wrote, "Throughout all past time, according to the records handed down from generation to generation, we find no trace of change either in the whole of the outermost heaven or in any one of its proper parts." Right from classical antiquity, astronomers and philosophers (there was no distinction between the two until the modern age) divided the night sky into two categories—first, the fixed stars that appear to rise and set but which retain their relative arrangements over time, and second, "wandering stars," which included the planets, the sun, and the moon.[4]

Fixed stars were also part of the powerful symbolism in ancient Hellenistic and Indian astrological traditions. Astrology, intimately propelled by observations of the night sky, in many ways paved the road to the development of the modern scientific discipline of astronomy. One of the earliest documents to deal with stars and constellations is a catalog found in a Latin astrological compendium called the *Liber Hermetis*. This star catalog may date back as far as 130 BCE. In any case, it seems to predate Ptolemy (c. 150 CE), although it does mention the names of many of the same stars he enumerated later in his *Almagest*. Ptolemy listed 1,020 new fixed stars in addition to those in the *Liber Hermetis,* and they became important to the Hellenistic tradition.[5]

The concept of an unchanging heaven has inspired many poets as a metaphor for permanence and constancy in a world that is ephemeral and changing. This idea that there was a fixed realm, however remote and unreachable, provided the human psyche with a sense of stability. No matter what else came and went, the stars were durable and silent witnesses to the short-lived drama of the human lifespan. The eternal backdrop of order reiterated a preordained divine origin for the cosmos. Fixity not only held sway in the human imagination as a fact but also offered a way to anchor human experience. In literature, we see particularly evocative examples. In Dante Alighieri's fourteenth-century allegorical poem *The Divine Comedy,* the eighth concentric sphere, depicting paradise, is that of the fixed stars, as postulated by Ptolemy.

William Shakespeare's lifetime (1564–1616) coincided with those of Giordano Bruno (1548–1600), Galileo Galilei (1564–1642), Tycho Brahe (1546–1601), and Johannes Kepler (1571–1630), all natural philosophers—or, as I will call them anachronistically, early scientists. Shakespeare was greatly influenced by their discoveries. Galileo's refinement of the telescope was radically extending vision outward and transforming our knowledge of the celestial sphere. In terms of world view, though, Ptolemy's geocentric theory, as propounded in his *Almagest,* still held sway. This was the dawn of the age of astronomy—and Shakespeare made frequent use of astronomy in his writings. He often evoked the fixed stars in his plays and his sonnets. In "Sonnet 21" fixed stars denote the steadfastness of love:

> O! Let me, true in love, but truly write,
> And then believe me, my love is as fair
> As any mother's child, though not so bright
> As those gold candles fixed in heaven's air.[6]

The symbolism of fixed stars remained in fashion for the English romantic poets. Percy Bysshe Shelley wrote in the fifth canto of his 1813 *Queen Mab,*

> How many a Newton, to whose passive ken
> Those mighty spheres that gem infinity
> Were only specks of tinsel, fixed in heaven
> To light the midnights of his native town.[7]

In this poem, Queen Mab and the Spirit of Ianthe "make a 'sublime' ascent from the earth in a magic car" to reveal "humanity's future paradise." An unusual feature of this poem is the length of Shelley's accompanying notes—a copious ninety-three pages added to the poem's own eighty-six. This is an example of Shelley's engagement with science, in which he supports his visions of poetry and the prophetic elements he employs with recently discovered and derived scientific claims—a marked shift from Poe's attempt to convince merely by suggestion.[8]

Though not a poet and now writing in the twentieth century, Einstein was no less entranced by the fixed stars. For evidence, we need look no further than his 1917 paper on the theory of cosmology, where he outlines the implications of a now famous new theory of gravity, general relativity: "Kosmologische Betrachtungen zur allgemeinen Relativitatstheorie" (On the cosmological problem of the general theory of relativity).[9] Einstein's so-called field equations of general relativity explain how matter and energy both generate gravity, and how gravity in turn impacts the shape of space and time. They also include a term, the cosmological constant, denoted by the Greek letter *lambda*. Lambda, a counterforce that opposes the attractive nature of gravity, in Einstein's formulation ensured that the stars and the nebulae (as galaxies were then known) remained fixed in the sky. Einstein argued that the value of lambda could be chosen to sustain this delicate balance, which would assure an unchanging universe with a fixed size. He designed this term very cleverly to safeguard all other observations that ratify his general relativity theory. Lambda's repulsive effects would produce negligible observational consequences on the scale of our solar system and would manifest only over the largest cosmic distances. These distances were well beyond observational reach at the time.

In the conclusion of his paper, Einstein admits, "That [lambda] term is necessary only for the purpose of making possible a quasi-static distribution of matter, as required by the fact of the small velocities of the stars." In other words, he did not have an explanation for why and how this lambda term arose. He justified his doctoring by claiming a need to be consistent with the small peculiar velocities or apparent motions of nearby stars with respect to a more distant frame of reference. But adding the extra term was not only a way of revising an equation to better represent the theory. Einstein's rationale for the modification showed a continuation of cultural tradition, a deep-seated belief in a static universe.[10]

In the static universe, Einstein believed that he had found the only possible solution to his field equations. But in 1917 the Dutch physicist Willem de Sitter showed that another solution existed. It described an empty universe, bereft of matter. De Sitter proposed a new model for the universe based on Einstein's cosmological theory, which he modestly and deferentially referred to as the "Solution B" to Einstein's "Solution A." The geometry of space, which is a key feature in Einstein's theory of general relativity, does not vary in time in Einstein's Solution A or in De Sitter's new Solution B. De Sitter, however, boldly assumed that the matter content of the universe was insignificant compared to the strength of Einstein's cosmological constant. In his solution, since there is no matter in the universe, its fate is dictated entirely by Einstein's fudge—the cosmological constant term. De Sitter's Solution B has two startling implications: measurements of time depend on an observer's location in the universe, and nebulae move bizarrely—they rapidly disperse away from one another, propelled purely by the intense repulsion of the dominant cosmological constant term, and thereby overrule gravity.[11]

De Sitter keenly followed developments in observational astronomy and knew about the astronomer Vesto Melvin Slipher's observations of retreating nebulae published in 1913. Einstein did not keep abreast of observational developments in astronomy. In his 1917 paper, De Sitter reported the observations of a few nebulae that were hurtling away at several hundred kilometers per second. These observations aligned with his prediction and thus supported Solution B, he argued. This did not convince Einstein or others. They deemed De Sitter's model universe absurd, because it contained no matter at all! Even though his Solution B was rejected, De Sitter's work was foundational, because he opened up a radical possibility—that of treating time in Einstein's equations as variable. De Sitter had initiated and advanced the concept of an evolving universe; however, what was needed were solutions that were more in line with the real universe—one that was visibly filled with galaxies and not empty.

Once De Sitter opened the door to considering a universe changing in time, it did not take long for the idea to percolate and for others to explore the possibility further. One such explorer was the Russian scientist Alexander Friedmann, who in 1922 began to investigate the behavior of solutions to the field equations describing a universe that contained matter and changed in time—that is, dynamical models of the cosmos with matter. He discarded both Einstein and De Sitter's ideas and with these new assumptions found yet other solutions, time-varying solutions that satisfied the field equations. In his model the universe was initially very dense, but it expanded and diluted progressively over time. Einstein read Friedmann's paper but dismissed the work summarily, as he vehemently disagreed with Friedmann's calculations. Partly as a consequence of this rebuttal, the paper never garnered wide readership.

Besides, Friedmann died young just three years later, at the age of thirty-seven. Without a strong advocate, his idea remained ignored.

In fact, Einstein was displeased with the solutions presented by both De Sitter and Friedmann, but for slightly different reasons. De Sitter's solution was absurd to him, as it postulated an empty universe. And Friedmann's solution contradicted Einstein's intuitive attachment to a static universe. In response, Einstein published several hastily written (and wrong) papers claiming to find errors in both of their calculations. But when the errors in his own responses were exposed, he admitted that these were possible solutions, even though he remained unconvinced. So even a figure whom many see as the iconic scientist, despite relying on reason and logic, held on to beliefs that had no rational basis. Einstein's conviction that the universe had to be static was unflinching until the observational evidence was insurmountable.

Theory and observation in astronomy were on parallel tracks until this point, and suddenly there appeared on the scene a European clergyman who made them intersect. The unassuming young Belgian priest and physicist Georges Lemaître made the vital connection between these speculative theoretical solutions and empirical data that finally spurred the acceptance of Hubble's findings. While at Harvard College Observatory in 1924–25, Lemaître realized the far-reaching implications of putting the theory and the data together. He attended the annual meeting of the American Astronomical Society in Washington DC in 1925, where he heard about Hubble's first major discovery of the existence of other galaxies, distinct from our own. He was also aware of the Indiana farm boy–turned–astronomer Slipher and his results of the retreating nebulae. Lemaître noted that these two observations put together implied that the universe was expanding. It clicked for him immediately. An observational test—to provide the solid proof needed to ratify the theoretical solution of an expanding universe—was slowly emerging. After returning home to Leuven, Belgium, he worked out the model for a universe in motion, similar to Friedmann's earlier work, even though he was completely unaware of Friedmann's ideas. The prescient Lemaître was two steps ahead of everyone else and immediately began to ponder the implications of Hubble's findings and the potential use of these newfound galaxies to test the properties of the universe. He was keen to test if our universe as observed was in consonance with general relativity. In a 1927 paper, he predicted that the speed at which nebulae hurtle away from us is proportional to their distance from us and concluded, "The receding velocities of the extra-galactic nebulae are a cosmical effect of the expansion of the universe."[12] The linear relation claiming that the speed of these receding nebulae is proportional to their distance from us was a brand-new result, one that Friedmann had not noted earlier. Now there was a clear prediction of theoretical solutions that could be directly compared to astronomical observations. Lemaître did not know of Friedmann's purely theoretical calculation, because that paper had sunk into oblivion. Unfortunately, Lemaître published his own breakthrough idea in French in an obscure journal, *Annals of the Scientific Society of Brussels*. So even while he was in Cambridge in 1928 in the company of great minds like Arthur Stanley Eddington, an intellectual giant of the British astrophysics establishment, Lemaître was unable to garner attention for his work. The theoretical idea of a universe in motion was published in the scientific literature by 1928, but no one saw it or was swayed by it.

We now jump back to 1912 and the world of observational astronomy, the backdrop to these theoretical developments. Observational astronomers had found hints much earlier that the universe was in motion. As mentioned previously, the first clues came from Slipher's measurements of the speeds of nebulae, made with the twenty-four-inch telescope at the Lowell Observatory in Arizona in 1912. The major instrumental advance at this time was the use of photographic plates on telescopes to capture images of faint astronomical objects. Although, an image of a celestial object was successfully taken in 1840, it would take much longer for the techniques to mature. It took another five decades to image dim stars and faint nebulae that could not be seen by eye. By the early 1900s, observers regularly attached cameras and other instrumentation—such as spectrographs, which reveal the chemical signature of light by displaying its component frequencies—to telescopes, enabling the detailed study of the light emitted from objects in the night sky. Researchers pointed telescopes at specific targets and collected light over long exposure times. This data registered as dark specks (as a negative) on a photographic plate that recorded both their position and their brightness.

This revolutionary technology gave astronomers the ability to make permanent records with long exposures that brought faint distant objects into view. Images on the plate meant that astronomers had lasting proof of what they saw. Photographic plates allowed researchers to analyze their observations and to make measurements of the properties of the objects in the photographic frame. Now, with this material evidence in hand, astronomers could revisit and study their photographic plates in the light of day. These plates also allowed easy transportation and sharing of observations. Most important, because measurements of an object's brightness could be made and calibrated, quantitative studies of statistical samples became possible. In particular, time-variable phenomenon could now finally be discerned and studied with repeated observations. Quantitative evidence could now be extracted and documented from an objective source—photographic plates—unlike the trained yet potentially biased astronomer's eyes. While this may not sound like a major advance to us today, in a field like astronomy, where no controlled laboratory experiments can be performed, this was a breakthrough moment. Instruments that reduced the reliance on the subjective observer and registered data automatically were the closest that cosmology was ever going to get to experimental data.

Photographic plates were, without a doubt, the crucial instruments that catalyzed the discovery of the expanding universe. They captured the first enduring images of the night sky that could be used, stored, and reused for research and analysis. They rendered the night sky in detail, which facilitated more thorough study of individual objects.

A photographic plate, a precursor of photographic film, is a glass plate coated with a photo-sensitive emulsion composed of silver compounds on which an image can be recorded. Plates were in use in astronomy until the 1990s, because they were more stable than film, and did not bend or warp easily. Many of the famous astronomical surveys of the sky recorded data on plates. Before the invention of digital cameras, the workhorse of the astronomical community was the photographic plate.

Photography has a long and richly documented history, but what is of concern for our story is the role that it played in capturing images of the night sky. Photographs of astronomical objects began to be used for scientific purposes in the mid-nineteenth century. Long exposures were needed to photograph faint astronomical objects, and telescopes needed to be structurally stable, as well as moved constantly, to compensate for the earth's rotation. It was a technical challenge to keep the telescope focused on a fixed patch of sky over a long period. The inventor of the daguerreotype process, Louis-Jacques-Mandé Daguerre, took the first picture of the moon in 1839. However, it ended up looking like a piece of fuzz due to the difficulty of tracking during the long exposure. William Cranch Bond and John Adams Whipple took the first photograph of a star on July 16–17, 1850, with Harvard College Observatory's fifteen-inch telescope, which still sits on its own tower at 60 Garden Street in Cambridge at the Harvard Smithsonian Center for Astrophysics campus.

Then, in 1871, Richard Maddox invented the lightweight gelatin negative plate. By trial and error with various materials, he discovered that plates coated with cadmium bromide and silver nitrate fixed with gelatin were incredibly sensitive to light. It was John Herschel, son of the astronomer William Herschel, who produced the first photograph on a glass plate and coined the photography terms *positive* and *negative.* John Herschel was a remarkable scientist in his own right, and his influential essay "Preliminary Discourse on the Study of Natural Philosophy" was published in 1831 in Dionysius Lardner's *Cabinet Cyclopaedia.* It articulates the method of scientific investigation and inspired many scientists, including the naturalist Charles Darwin.[13] Herschel developed the technique of coating a glass plate on one side with a gelatin emulsion containing small crystals of the light-sensitive compound silver halide. The size of the crystals determines the sensitivity, contrast, and resolution of the image. The emulsion slowly darkens when exposed to light, capturing the gradations in the image and leaving an imprint.

By early 1900, photographic plates were routinely being used to take astronomical images. Intense manual work was needed to extract the information, for which Edward Pickering, the director of the Harvard College Observatory, employed a team of women researchers, including Henrietta Swan Leavitt—whose work was crucial to Hubble's mission—at a salary of 25–30 cents an hour. Pickering recruited Leavitt and other women with college degrees to work on his ambitious survey to catalog and accurately measure the brightness and color of every star in the sky. In the 1950s, the Watson Scientific Computing Laboratory, established at Co-

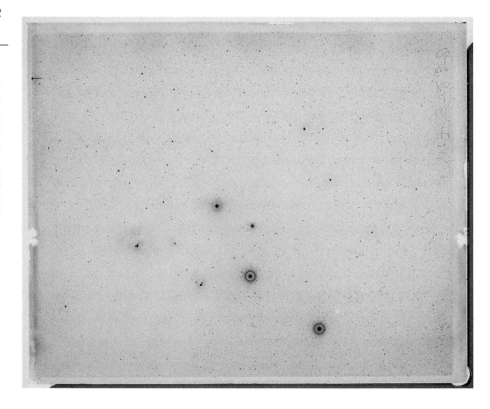

A portion of the photographic plate mf37250, an image of the multiple star system Rho Ophiucus in the constellation Ophiucus, taken on May 30, 1948. Courtesy of Harvard College Observatory.

lumbia University in collaboration with IBM, pioneered a new and automatic way to measure astronomical plates, so machines finally replaced these "human computers." Automating the plate measurement process helped extract data from generations of large area sky surveys thereafter.

The power of telescopes and photographic plates lay in how they suddenly rendered the invisible visible, the ephemeral concrete, and the fleeting permanent. These developments extended our senses, enhanced objectivity, and refined how information could be turned into evidence. Astronomical observations became the engine of discovery providing evidence for cosmic phenomena.

It was Slipher, mentioned above, who discerned one of the early hints from data using these new observational tools, finding in 1912 that the Andromeda Nebula appeared to be rushing toward us at a rather dramatic speed of about three hundred kilometers (186 miles) per second, which is about a million kilometers (more than six hundred thousand miles) per hour. By 1914, by measuring their velocities, he found that several other nebulae were also peculiarly flying rapidly, but away from us. These speeds were impossible to conceive. The fastest car in the 1912 Indianapolis 500 rally, for example, clocked in at an average speed of only 129 kilometers (eighty miles) per hour. Part of the puzzlement of these measurements was how entirely beyond human grasp these tremendous speeds were.

External galaxies (which were known as extragalactic nebulae at the time) were known to be aggregates of stars that were either too dim or packed too close together to be distinguished individually by the naked eye. Today we know that galaxies like our Milky Way contain about a hundred billion stars, gas, and dust and that there are many billions of galaxies aside from our own in the universe. At Slipher's time, the precise distance to nebulae was unknown, and one of the questions that astronomers debated frequently was whether these distant nebulae were inside our galaxy or if they were island universes beyond our galaxy. The inferred extent of the universe then, like today, was defined by the visible edge—how far out we can see with the best available instruments. As a concept, the idea of nebulae as aggregates of stars isolated in space was not new. The English astronomer Thomas Wright (1711–86) had formulated it as early as 1750. Wright, preoccupied lifelong with a desire to reconcile his religious and scientific views, conceived of these astronomical entities within the framework of a cosmotheological vision. Prior to distance measurements being made, it was conjectured that any sample of the universe is much like any other and therefore if all stars are assumed to be roughly as bright as the sun, the fainter ones appear so because they are simply farther away. Their distances could therefore be estimated by comparison with the brightness of the sun. The creative Wright, however, speculated beyond the Milky Way. He imagined that there could be nebulae well beyond our habitat.

A contemporary of Wright's, the philosopher Immanuel Kant, strongly supported this assertion of the existence of many external nebulae beyond our own galaxy, referring to them as island universes. In his *Allgemeine Naturgeschichte und Theorie des Himmels* (*Universal Natural History and Theory of the Heavens*), published in 1755, Kant wrote (as translated), "We see scattered through space out to infinite distances, there exist similar systems of stars [nebulous stars, nebulae], and that creation, in the whole extent of its infinite grandeur, is everywhere organized into systems whose members are in relation with one another. . . . A vast field lies open to discoveries, and observations alone will give the key."[14]

In his Silliman Memorial Lectures delivered at Yale University in 1935, published under the title *The Realm of the Nebulae,* Hubble described Wright's speculation: "A single stellar system, isolated in the universe, did not satisfy his philosophical mind. He imagined other, similar systems, and as visible evidence of their existence, referred to the mysterious clouds called nebulae."[15]

Astronomers had started studying nebulae, and Slipher had, by 1914, measured the speeds of thirteen of them, using the change in wavelength of light that results from relative motion of the source with respect to us. Just as the sound of the siren of an approaching ambulance is shriller (has a higher pitch) when it is moving toward us, light emitted from a body moving toward us is shifted in frequency and wavelength toward the blue end of the spectrum, or blueshifted. Conversely, when the object emitting light is moving away from us, the wavelength shifts to the redder end and is measured as a redshift. This phenomenon is known as the Doppler effect. Using this effect, whereby the wavelength stretch or compression of the light that we receive on earth reflects the motion of the object, Slipher estimated that the typical nebula was moving away from us at a speed of approximately six hundred kilometers (373 miles) per second, which is significantly higher than the speed of any known object in the sky. Over the next eight years he amassed data for about forty such nebulae and found that they all seemed to be consistently receding, with the exception of Andromeda. Astronomers, including Hubble, pondered Slipher's results, and even eminent theorists, such as Eddington, were baffled by these large speeds. These measurements were confusing to interpret, but it was acknowledged that they were important and required further study and understanding. At the time, the fact that these nebulae were extragalactic was not recognized, as the key unknown was their distance from us.

In 1912, Leavitt at the Harvard College Observatory provided the critical step forward. Pickering, the observatory's director, was interested in covering a large swath of the night sky and therefore in the exploration of the statistics of astronomical objects, while Slipher looked at individual galaxies longer and deeper. Pickering's all-female labor force pored over the accumulating photographic survey image plates with magnifying glasses, meticulously making measurements. Leavitt and the rest of the team were responsible for calculating the positions and brightness of stars, which appeared as tiny black specks (as the negative) on glass photographic plates. By this time, photographic plates were quite sensitive, so each one contained the image of more than a thousand stars as dark specks. Pickering's army of women, his human computers, performed these extremely tedious tasks of measuring and recording the properties of the brightest stars in these crowded photographic plates.[16]

Astronomers realized that if the true brightness of a star were known, then its faint appearance could be used to directly measure its distance from us. For example, once we know that the brightness of a shining bulb is sixty watts, if it is observed to be four times fainter

Pickering's "human computers" at Harvard College Observatory: group photograph of the women employed to analyze astronomical data, taken in 1913. From the Harvard University Archives UAV 630.271 (E4116).

than that, then we can infer that it is twice as far away as a bulb right overhead. Light sources need to be standardized to make relative comparisons. Leavitt found just such a set of stellar light bulbs with known wattage (so-called standard candles): Cepheid variable stars. Although it might seem counterintuitive to think of stars that vary as candidates for steady candles, there is a surprising regularity about their variation that allows their use as calibrators. The brightness of these stars varies in a regular and predictable fashion in cycles ranging from about a few days to a few months. Leavitt discovered that there is a correlation between the intrinsic brightness of a Cepheid (the true wattage) and its pulsing. Her work, looking for small changes in multiple photographic plates of the same region of the sky, was painstaking. Brighter stars appeared as larger dark specks.

She compared the sizes of these dark specks with those on a calibrated, standard "fly spanker plate"—she measured the brightness of several stars in each frame and checked for changes in the brightness of individual stars one by one. Having examined hundreds of plates, Leavitt was the undisputed expert in estimating the brightness of a star imprinted on a photographic plate. She looked for varying stars whose brightness changed in a regular manner over a fixed interval. To compare plates of the same region of the sky taken at different times, they had to be lined up with a positive photograph of the same patch of sky taken on another date. If the black and white specks on the negative and positive images did not match properly, Leavitt identified the star as a variable. Executing this search in a fiendish fashion, she reported the discovery of 1,777 new variable stars in the Magellanic Clouds in the Southern Sky in 1908. At the very end of an article published in the *Harvard College Observatory Circular*, she lists sixteen special stars (later identified as Cepheid variables) for which "the . . . brighter variables have the longer periods." As all these stars were in the same "cloud," or nebula, and therefore likely at roughly the same distance from the earth, she was able to conclude that their variation periods had to do with their light and not their distance. Brighter Cepheids had longer periods. Turning this around, Leavitt realized that she could estimate the distance to Cepheids. Since two Cepheids with the same luminosity have the same period, if one appears brighter than the other, it is definitely closer to us. The reason is straightforward—the apparent brightness diminishes with the square of the distance. A star twice as far away as another but of equal apparent brightness is four times as luminous. Leavitt's proposed method to measure distances therefore involved the following steps: measure a Cepheid's period and its apparent brightness, use the period-luminosity relationship to estimate its intrinsic brightness, and then compare the intrinsic to the apparent brightness to derive its distance from the earth. It was only

Leavitt who was in a position to discover Cepheids because, serving as a human computer, she saw more plates—and more stars—than any of her counterparts.[17]

Of course, for Leavitt's method to work, calibration was essential. A sample of nearby Cepheids whose distance could be measured by another independent technique, like parallax, was needed. Unfortunately, there was not a single detected Cepheid variable in the Milky Way galaxy close enough for its distance to be measured by stellar parallax. The only way forward at the time was to approach the problem statistically, using what is known as statistical parallax, involving a collection of Cepheids in the Milky Way whose combined slow motions across the sky were known. The hunt for Cepheids was on. The "golden boy" of Mount Wilson, the astronomer Harlow Shapley, began to look for Cepheids at various locations in our galaxy and, on finding them, concluded (incorrectly) in 1920 that all the nebulae lay within our galaxy. Later he used the Cepheids in globular clusters to show that the Milky Way was an enormous system of stars vastly larger than anyone had previously supposed. This was the first successful application of the Cepheid variable distance method, and Shapley used it to estimate the size of our galaxy.[18]

To extend the reach of Leavitt's technique, astronomers needed to search farther for this particular class of star, in external nebulae, to obtain an estimate of distance. Leavitt's Cepheid method rapidly became the standard cosmic yardstick. This was the backdrop against which the ambitious and energetic young Hubble arrived in California to look through the most powerful telescope in the world. Even before finishing his PhD at the University of Chicago in 1917 with a thesis titled "Photographic Investigations of Faint Nebulae," Hubble was offered a job—a research position at the Mount Wilson Observatory in Southern California. He deferred it for one year, to serve in World War I. Although he delayed his return in the end by a couple of years, the job was still waiting for him when he left the military with the rank of major. At this time, institutions on the West Coast dominated observational astronomy. The Mount Wilson Observatory, operated by the Carnegie Institution of Washington, and the University of California's Lick Observatory on Mount Hamilton were at the frontier, with the best telescopes and facilities. The most powerful telescopes were the thirty-six-inch Crossley reflector at Lick and the sixty-inch reflector on Mount Wilson. Hubble's timing could not have been more opportune, because when he returned to astronomy, the new hundred-inch Hooker telescope, soon to be the largest in the world, was just weeks from completion for Mount Wilson.[19] All of thirty years old, bursting with ambition, Hubble had access to the best available instrument. The stars aligned superbly, so to speak, because his two potential competitors, Heber Curtis and Shapley, moved away from California just prior to his arrival and soon after, respectively, leaving the frontier wide open for him. Curtis and Shapley were mired in the debate over whether the nebulae were island universes like our own but distant from ours or just clumps of stars that were part and parcel of our galaxy. They disagreed on the size of the Milky Way and on the distance to the nebulae. Curtis did not accept Leavitt's period-luminosity equation, while Shapley successfully used it to determine the size of the Milky Way. Curtis, who wanted to photograph spiral galaxies at frequent intervals to look for novae and variables, as Hubble eventually did, took himself out of the fray by leaving Lick in 1920 to become the director of the Allegheny Observatory in Pennsylvania, where he could not pursue this research program. However, a lifelong rivalry between Hubble and Shapley began during the two years that they overlapped at Mount Wilson. Shapley then moved to Cambridge, Massachusetts, and after a trial period became the director of the Harvard College Observatory. Despite detesting each other, they remained in constant correspondence, as the strategic and politically shrewd Hubble always shared his ideas and discoveries with the powerful and influential Shapley.[20]

Hubble decided to tackle the challenge of measuring distances to nebulae. Using Leavitt's method, he began to look for more Cepheids. This took many long, lonely nights of observing with the help of an assistant, sitting still on the platform next to the instrument mounted on the hundred-inch mirror and tracking stars. Hubble dressed warmly, and once the telescope clicked into place and pointed to where he intended, he would light his pipe and sit hunched over, watching the heavens swing slowly by. On one of these celestial vigils, much to his excitement, he found one Cepheid in the Andromeda Nebula (our nearest neighboring galaxy) and successfully measured its distance from earth to be nine hundred thousand light-years, about three times the size of our galaxy as estimated by Shapley using the same techniques on nearby cosmic objects.[21]

This demonstrated clearly that the Andromeda Nebula was an external galaxy well outside our own. Hubble soldiered on, measuring distances to several other nebulae, and showed that they all lay well beyond the Milky Way. Thus, he established the existence of truly extragalactic nebulae, which skirted the heavens beyond our galaxy. Hubble worked diligently and patiently on his program of imaging spiral nebulae. His first major discovery and claim to fame was the determination of reliable distances using Cepheids to the nebulae NGC 6822 (detecting eleven variables) and Andromeda (now detecting eleven additional variables), showing that they were seven hundred thousand and one million light-years away, respectively. These massive distances, much larger than the estimated size of our galaxy, implied that they were external to the Milky Way. Hubble made his mark by settling the rancorous debate between Curtis and Shapley once and for all in 1925.

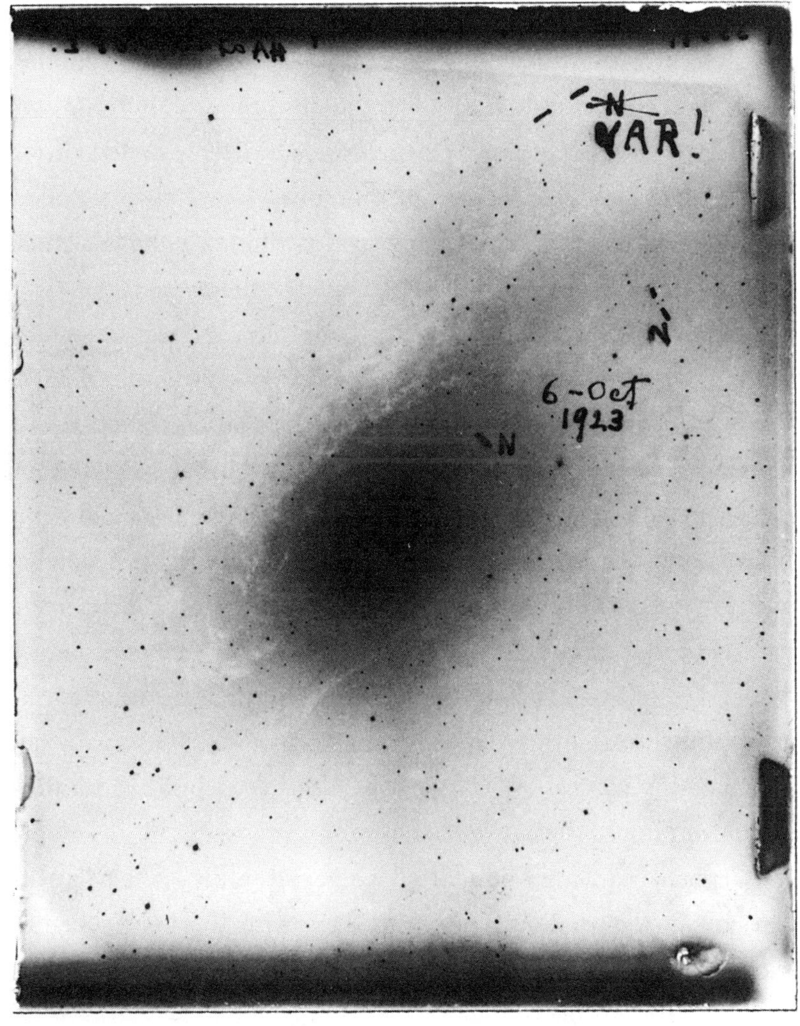

Image of Hubble's M31 plate, which marks the position of the first Cepheid variable, a type of star used to determine distance that Hubble discovered in the Andromeda Galaxy (M31). He originally thought that this was a nova and marked it with an *N*. On realizing that it was a variable, he wrote "VAR!" on the image. Courtesy of the Observatories of the Carnegie Institution of Washington.

Our galaxy was by no means unique, simply one of many that dot the universe. At this point, not only was the earth just another planet in the solar system, revolving around the sun, but our own Milky Way was also one of many galaxies. The deeply rooted notion that we occupy a special place in the universe was now dispelled. But this was just the start of yet another, more forceful undoing that radically reconstituted our knowledge map of the cosmos.

Hubble's reputation was firmly established in the astronomy community after this discovery. Bringing his lawyerly training to bear, he was an extremely careful and cautious observer who always strategized how to present his results—seldom in a hurry. He took every precaution, worked hard to woo potential enemies and competitors, and strove to make an airtight case to convince skeptics of the veracity of his observations. Besides, he never attempted to interpret his observations through the lens of any specific theoretical model—he left that task to the theorists and others. Continuing to determine the distances to several other galaxies, Hubble finally saw the emerging pattern of how galaxies retreated depending on their distance from us.

Cepheids could not be spotted easily and studied if they lay beyond about five million light-years, even with the unprecedented reach of the hundred-inch telescope. Hubble pushed out farther by using classes of some of the brightest stars—the O and B stars—as standard candles. With this methodology in hand and with the best possible instrument available to him, he eventually turned his attention to Slipher's sample of nebulae that were hurtling away from us. Although Hubble had access to Slipher's data, he was unaware of both Friedmann's and Lemaître's theoretical papers. It is unlikely that Hubble had even encountered the idea of an expanding universe or knew how to interpret his data in light of such a theory. Theory beckoned in the direction of a radical new picture of the universe, as a dynamic entity. This was a huge philosophical and intellectual leap, as it implied that somehow space itself was stretching, a concept that is very hard to grasp.

Aided by the sharpest instruments available at the time, working at the world's best astronomical observatory, Hubble did what no one else could. In 1929, he presented his data and Slipher's and showed that there appeared to be a linear correlation between the velocity at which a nebula was receding and its distance from us. Thus, a nebula twice as far away hurtles away at twice the speed. This relation is now known as Hubble's law. The constant of proportionality that relates speed and distance is referred to as Hubble's constant. Although a mild trend was there in the data that Hubble presented in an initial sample of twenty-four nebulae, it was by no means persuasive. Buoyed with confidence from his earlier success and no longer the overcautious scientist, he boldly claimed that the data in the graph in his paper indicated the presence of such a correlation. He did make a sort of leap of faith by postulating this relationship, and despite his ambition and zeal, after making this remarkable discovery, Hubble and the astronomical community took several years to realize its significance and import. It was only on extending the measurements to galaxies at distances even greater than those reported in his first paper that Hubble could make a convincing case for a linear relation, which he did in a paper written with Milton Humason in 1931. It took the confluence of Lemaître's theoretical models and Hubble and Humason's observations for astronomers to recognize how radical Hubble's findings were. Of course, in all of this Hubble himself remained skeptical of the expanding universe interpretation, despite being the one who uncovered the definitive evidence for it!

The onset of the complete reshaping of our world view was well under way by 1931. Hubble's findings portended what came next—increasingly dramatic discoveries that upended our notion of a stable and fixed cosmos. We now had the first glimpse of our restless universe.

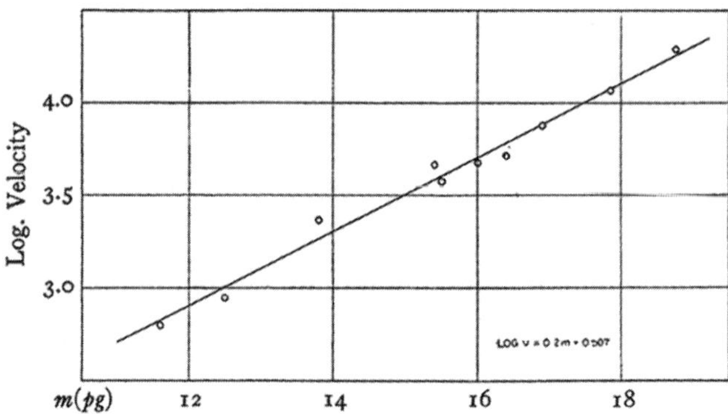

FIG. 4.—Correlation between the quantities actually observed in deriving the velocity-distance relation. Each point represents the mean of the logarithms of the observed red-shifts (expressed on a scale of velocities) for a cluster or group of nebulae, as a function of the mean or most frequent apparent photographic magnitude.

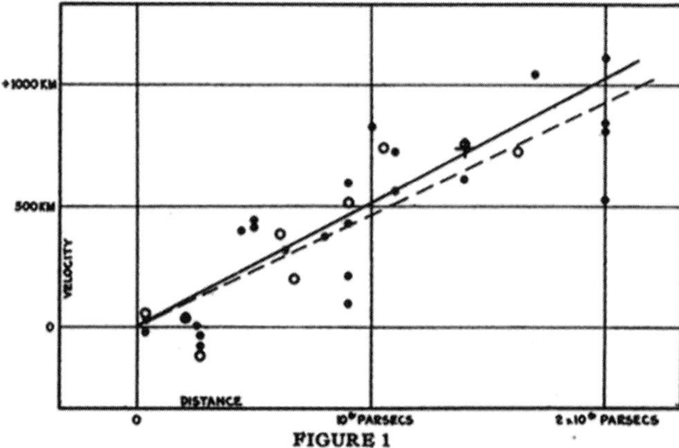

FIGURE 1

Velocity-Distance Relation among Extra-Galactic Nebulae.

Radial velocities, corrected for solar motion, are plotted against distances estimated from involved stars and mean luminosities of nebulae in a cluster. The black discs and full line represent the solution for solar motion using the nebulae individually; the circles and broken line represent the solution combining the nebulae into groups; the cross represents the mean velocity corresponding to the mean distance of 22 nebulae whose distances could not be estimated individually.

Top: Hubble and Humason's data published in the paper "The Velocity-Distance Relation among Extra-galactic Nebulae," *Astrophysical Journal* 74 (1931): 43, fig. 4. © AAS. Reproduced with permission.

Bottom: Hubble's point justifying the postulate of a linear line, from "A Relation between Distance and Radial Velocity among Extra Galactic Nebulae," *Proceedings of the National Academy of Sciences* 15, no. 3 (March 15, 1929): 172, fig. 1.

To understand how theory and observations collided and meshed, we need to return to Lemaître's mathematical solution to Einstein's equations. Lemaître's theoretical prediction that the recession velocities of nebulae are proportional to their distance holds only in a universe where the mass is evenly distributed, because there is then uniform expansion in all directions simultaneously. Lemaître's solution is not valid if there are regions in the universe where mass is distributed in lumps—his model requires the universe to be featureless and fairly homogenous. The data that Hubble and Slipher had collected extended only as far as about six million light-years and showed that space was replete with galaxies, hardly homogenous. Our current understanding of the distribution of matter in the universe reveals that the assumption of homogeneity is valid only on much, much larger scales than those to which Hubble initially ventured out. On the enormous scales that we have access to today, the granularity produced by individual galaxies begins to smear out, just as the bumpy structures of individual cells in our skin appear smooth.

According to Lemaître, on smaller scales—those on which Hubble and Slipher made their measurements—the universe is not meant to be uniform, and a linear relationship is not expected to hold between recession velocity and distance. Unaware of Lemaître's model predictions when he wrote his paper in 1929, Hubble simply made a bold guess that turned out to be true. In fact, it was only after extending their reach to a distance of one hundred million light-years that Hubble and Humason found much more convincing evidence for a linear relationship. The data in their joint 1931 paper justifies the claim of this relationship, as they had just reached the edge and sampled the regime where homogeneity kicks in and Lemaître's solution is valid. Hubble did not interpret the linear relation as a signature and consequence of a homogenous and expanding universe. All he knew was that his findings had cosmological relevance. Realizing the far-reaching implications of his results was a task he left to the theorists, primarily Lemaître with support from Eddington. In their 1931 paper, Humason and Hubble mention De Sitter's theoretical model, albeit just in passing, but not Lemaître's. Incidentally, Slipher had secured almost all the reported redshifts in Hubble's 1929 paper, yet Hubble does not acknowledge him. Rivalries are mirrored in these seemingly small oversights in giving explicit credit. This unfortunate practice, motivated by competitiveness, the rush to publish first, and inadvertent or occasionally intentional refusal to acknowledge the work of others in the service of personal ambition, is alas still prevalent today. Such lapses are a consequence of the race to be the first to report new discoveries and claim intellectual credit.

Hubble was the inveterate experimental scientist whose thinking was entirely driven by data. He was, however, acutely aware of the need for a theoretical framework, and in *The Realm of the Nebulae* he muses, "Observation and theory are woven together, and it is futile to attempt their complete separation. Observations always involve theory."[22]

As the observational buzz was building with news of the growing number of fleeing nebulae, Lemaître decided to give one final push to his theoretical work. He sent Eddington a second copy of his 1927 paper after Hubble published the data in 1929. Eddington simply connected the dots, strongly encouraged Lemaître to republish his paper in English in the well-regarded and widely read *Monthly Notices of the Royal Astronomical Society,* and personally promoted the model. His advocacy forced Einstein to finally pay attention. Earlier, in 1927, Einstein had been brutal in his criticism of Lemaître's paper. "Your calculations are correct," he had said, "but your physical insight is abominable."[23]

At this point Einstein also recalled Friedmann's paper that he had criticized incorrectly and turned the tide in favor of the expanding universe model. In fact, after Hubble's seminar in the library at Mount Wilson in 1931, Einstein made a public announcement that he had been wrong about the static universe and that there was no use for the extra term that he had added to his equations. The journalist George Gray, writing soon after this event in the *Atlantic,* described the discovery as "a radically new picture of the cosmos—a universe in expansion, a vast bubble blowing, distending, scattering, thinning out into gossamer, losing itself."[24]

A new universe had been unveiled, one that had a beginning in time and continued expanding thereafter. A September 7, 1932, *New York Times* report from James Stokley, the associate director of the Franklin Institute Museum of Philadelphia, begins with this quote from Eddington: "The expanding universe now has a secure place in science."[25] Lemaître had provided an elegant theoretical framework within which to interpret the expansion that Hubble had found in the observational data. Hubble himself struggled to accept this interpretation of his linear relation as evidence of an expanding universe. He makes this amply clear in the abstract of his 1936 paper "Effects of Redshift on the Distribution of Nebulae": "The high density suggests that the expanding models are a forced interpretation of the data."[26]

Yet the story of Einstein and Hubble does not merely rest on the role of evidence and interpretation in settling the question of the fate of the universe; it also shows that personal beliefs can be hard to dislodge, causing persistent resistance to ideas even from those who find evidence for them. Hubble's discomfort arose from the interpretation of the data. He was unsure about whether the measured redshifts for the distant extragalactic nebulae were actually velocity shifts, as it is this identification that leads to the model interpretation of a homogenous expanding universe. For him, the assumption that redshifts are *not* velocity shifts was a more economical inference in the absence of any other satisfactory and compelling explanation in his mind. He felt that a choice was being made between accepting a small-scale universe and accepting a potentially new principle of physics. He really preferred the conclusion that his observationally measured systematic redshift effect was probably due not to a true Friedmann-Lemaître expansion but driven by an as yet undiscovered fundamental principle of nature.

Einstein's resistance, on the other hand, had deep roots. New evidence that demonstrates the extent of his reluctance to give up on the notion of a steady, non-evolving universe has recently come to light. Although he publicly accepted the idea of the expanding universe, a previously unknown manuscript, discovered among his papers in the archives of Hebrew University in Jerusalem in 2013, suggests that he privately struggled to resurrect a static universe even after that fateful seminar at Mount Wilson to which Hubble had invited him. In this manuscript, dated 1931, Einstein explored a model in which the mean density of the universe is held fixed by some process that continually generates matter from empty space. In this model, a static universe merely appears dynamic.[27]

Einstein investigated a fix to compensate for the expansion of the universe. The model expounded in this handwritten, four-page draft is very different from the many others that he had explored before. This newly discovered paper reveals his toying with and preempting the steady state cosmological model developed in the 1950s by Fred Hoyle, Hermann Bondi, and Thomas Gold. The discoverers of this unpublished draft, Cormac O'Raifeartaigh, Brendan McCann, Werner Nahm, and Simon Mitton, found that Einstein's calculation contains a fatal mathematical flaw, which is likely why he abandoned it. He appears to have grappled with steady state and evolving models of the cosmos well before, in fact decades prior to, the rest of the community. This unpublished model is outlined in the signed draft, titled "Zum kosmologischen Problem" (On the cosmological problem), which was previously assumed to be an early version of another paper. Einstein completely discarded this calculation; it is not referenced in any of the subsequent papers on cosmological models. In this manuscript, he constructs this model from first principles: the cosmological constant is retained, and there is no reference to either Friedmann's analysis or to any of Einstein's own evolving models that were published earlier in 1931. What is intriguing about Einstein is that while he enthusiastically wrote two papers with De Sitter in 1931 and 1932 on expanding universe models, he was still secretly playing with this idea of a steady state model. He was desperately trying to recover fixity. There is also no reference in this manuscript to the problem of cosmic origins, which was one of the main reasons for Einstein's displeasure with Lemaître's solution. So it appears it was not the dislike of that technicality—the awkwardness of having to explain the beginning—that motivated him to persist in the search for a steady universe. Perhaps this was his last-ditch effort to recover a static universe, even though it looked highly unlikely.

Although the majority of the astronomical community and the public were persuaded of the expansion of the universe by the end of the 1930s, even after the heavyweights Einstein and Eddington publically lent their support to it, there were some in the scientific community, including Hubble, who were still skeptical of this cosmological interpretation. In fact, Hubble retained his belief that the expansion might not be real right up to his final paper, which he gave as the George Darwin Lecture of the Royal Astronomical Society in May 1953, just four months before he died. Extrapolating Lemaître's solution back in time implied that there was a clear beginning of both space and time. The need for a starting point made several cosmologists uncomfortable. Friedmann and Lemaître's model of an expanding universe meant that earlier, the universe had been not only smaller but also denser. This, of course, inevitably led to the profound question of its origin. Lemaître postulated that the universe might have originated with a primeval explosion that subsequently generated the expansion. This implied that the universe had a provenance, an instant when it all began. While there was strong support building for this big bang model, it had some problems and open questions that still needed to be settled involving the formation of chemical elements and the age of the universe.

Despite the revolution caused by Hubble's data (and the debt he owed to Leavitt), neither of them was awarded the Nobel Prize. The astronomy community celebrated Hubble, and he won many awards and medals, however, he spent much of his later years advocating for astronomy to be included as a sub-field of physics. His intent was to canvass for astronomers like himself to be considered for the Nobel Prize. This unfortunately did not happen during Hubble's lifetime. The Nobel Prize Committee though eventually decided to expand the remit of the Physics Prize to include astronomy. In 1925 Gösta Mittag-Leffler of the Swedish Academy wrote to Leavitt, saying that he wanted to nominate her for the Nobel. He was unaware that she had been dead for three years.

Although the outbreak of World War II limited scientific progress in cosmology in the 1940s because of the redirection of scientific resources to the war effort, it also led to fundamental technological advances that reshaped the field in unanticipated ways. The development of new tools altered the kinds of questions that could be asked and answered. Advances in the field of nuclear physics enabled a new set of calculations of the abundance of chemical elements that would be produced in the initial cosmological explosion. In 1946, the Ukraine-born American physicist George Gamow calculated how a primordial soup of particles might create the various chemical elements. Assuming an initial state of an infinitely hot and infinitely dense cosmic stew—predicted by Lemaître—comprising radiation and the subatomic particles electrons, protons, and neutrons, Gamow and his collaborators estimated the abundance of hydrogen and helium that would be produced in the infant universe. They used electronic digital computers, developed for bomb calculations. There was rising discontent with this hypothesis of an initial cosmic explosion, and there was a new challenge mounted from across the pond—that the universe might be in a steady state and unchanging, though not static. What spurred the detractors of the big bang model was Gamow's failure to predict the formation of elements beyond hydrogen and helium. We now know that hydrogen and helium account for 99 percent of the matter of the universe, but heavier elements like beryllium, boron, and iron do exist. At that time, their origin was unclear. The early universe and the hot cosmic explosion did not appear able to predict their existence. It is this failure to predict the synthesis of chemical elements that led Hoyle to coin the term "big bang," rather pejoratively, because he believed that "the big bang is an irrational process that cannot be described in scientific terms . . . nor challenged by an appeal to observation."[28]

According to physics folklore, the idea for the eternal, steady state universe came to the three Cambridge friends Hoyle, Bondi, and Gold in 1947 after watching a movie, a ghost story that circled back and ended just as it had begun. The scientists' friendship dated back to their joint work on radar during World War II. They were a powerful trio—Hoyle was an extremely intuitive, versatile thinker, Bondi a fine mathematician, and Gold a very creative and imaginative scientist. Hoyle recalled, "One tends to think of unchanging situations as being necessarily static. What the ghost-story film did sharply for all three of us was to remove this wrong notion. One can have unchanging situations that are dynamic, as for instance a smoothly flowing river." This inspired them to consider whether the universe could look the same even if it were continually expanding.[29]

The only way this could be was if matter were being continuously created. Thus new galaxies could form and repopulate the region left behind as the older galaxies drifted apart. This new model, the steady state universe, incorporated expansion but dispensed with the idea of a beginning and an end. The universe was eternal

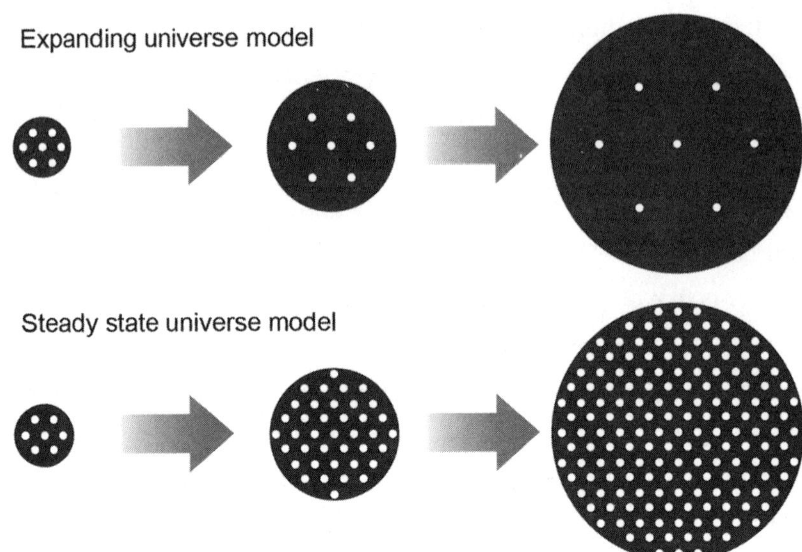

Schematic showing the expanding universe model and the steady state universe model. In the steady state model, the number of dots is preserved even as the universe expands.

according to the steady state model. To many cosmologists who were philosophically inclined, the steady state model that Hoyle, Bondi, and Gold proposed was appealing. First, as matter was being continually created, their universe did not get diluted, despite expansion. Second, this model circumvented the problematic question of the origin. Aside from failing to describe the origin of elements heavier than helium, the big bang model also predicted an age for the universe that was much lower than the known age of the solar system. These were clearly gaping holes in the big bang theory at the time.

So the proposal that the universe was homogenous in space and in time propelled the steady state model. Religious temperaments also interfered in this debate. In 1952, Pope Pius XII supported big bang cosmology as it was in consonance with the idea of a divine creator as postulated by the Church. Steady state theory, with no anchor in time, with no beginning or end of the universe, was seen to represent the atheist vision. Not all proponents of the steady state model were atheists, though—William H. McCrea, a leading exponent, was a committed Anglican. However, by and large, dispensing with the need for a beginning diminished the need for a purposeful creator, in consonance with atheist conceptions of the cosmos.

Astronomers in the United States also found the steady state model compelling, however they did not consider the matter settled. They felt the competing claims of the two theories ought to be resolved by observational tests and nothing else. The key findings in support of the big bang theory that ultimately decimated the steady state model came primarily from the measurements of the cosmic microwave background radiation and the age of the universe, which was determined to be comfortably older than our solar system, and from a deeper understanding of nucleosynthesis—the formation of chemical elements—which revealed how chemical elements heavier than helium were synthesized in the centers of stars and not in the early universe. Of course, there were attempts, principally by Hoyle, to resurrect the steady state model, but it was ultimately unsuccessful in explaining the growing wealth of observational data. For the steady state model, the death knell was provided by the 1965 discovery of microwave radiation, a relic of the big bang, an echo of the primordial hiss from the hot, dense beginning of the universe. The controversy between these two models raged fiercely for about twenty years, but increasingly, the steady state theory simply failed to explain the data that were now flowing from observations made with instruments that operated on many wavelengths—optical, radio, and microwaves. In his book *Cosmology and Controversy,* the historian Helge Kragh documents the similarities and tensions between the models and provides a detailed account of how this rivalry was finally settled.[30]

Besides, computational progress spurred by the invention of more sophisticated and faster computers during World War II under the Manhattan Project enabled a new suite of theoretical calculations. The complicated chemical networks and reaction rates needed to derive elemental abundances and stellar evolution were now calculable. All this newly developed instrumentation provided new predictions and laid the steady state model to rest. The critical data that confirmed the big bang model, though, was the discovery of the all-pervasive hiss of the microwave background, which we will discuss in chapter 5.

The demise of the steady state theory demonstrates the power of empirical observations and how accumulating evidence challenges or bolsters a theory. Notably, the steady state theory was falsifiable, because it made concrete predictions that allowed discrimination against the big bang model.

The discovery of the expanding universe and the establishment of the big bang model illustrate the powerful role that intellectually influential individuals play in the acceptance or rejection of new ideas. But they equally show that experimental evidence and data are the final arbiters. The powerful interplay of theory and observations in the 1920s and 1930s—the cooperative and fruitful interaction that unfixed the universe—also marked the birth of a new discipline, cosmology, the study of the properties of the universe and all that it contains. This new enterprise has since grown dramatically in influence and impact. In addition, one of the foundational assumptions of big bang cosmology that has been ratified with ample observational support since Hubble's initial discovery is the cosmological principle, which states that the universe is homogeneous, the same in all places, and isotropic, the same in all directions. With the extension of Hubble's data to farther reaches of the universe, it appears that on the largest scales, the universe obeys the cosmological principle. Another key assumption implicit in the interpretation of

Hubble's data that appears to hold true is that physical laws that we have discovered are valid in every patch of the universe, not only in our galaxy but also in every other galaxy near and far. Support for the big bang model—a universe with a dense, hot beginning that then began to expand at a constant clip—has continued to grow.

3

THE DARK CENTER

Black Holes Become Real

Black holes—the most massive and compact astronomical objects in the universe—provide rich fodder for our imaginations. Take for example Rumiko Takahashi's manga series *Inuyasha,* whose character Miroku carries the curse of a black hole in his hand. He inherits this curse from an ancestor, which ensures that anything material that comes in contact with his hand is instantaneously and violently pulled into a "wind tunnel" void. With every passing year, his black hole grows and threatens to draw in and kill Miroku himself.[1] This terrifying image of a mysterious object obliterating everything in its path is a common trope. After the financial meltdown of 2008, popular media described the black hole of Wall Street, and a scan of articles in the *New York Times* reveals the use of black holes to describe terrorism watch lists, intelligence reports on North Korea, and Mitt Romney's finances—alluding to a complete lack of information on these subjects.

As we have seen, the journey to acceptance of the concepts of the heliocentric solar system and the expanding universe was not linear—nor was the analogous path for black holes, from exotic mathematical entity to accepted theory and popular culture cameo. It was not the peculiar properties of an astronomical object that first inspired the term "black hole" but rather a place—an infamous prison. This was the scene of a gruesome incident on June 20, 1756. Siraj-ud-Daulah, the nawab, or ruler, of Bengal at that time, captured Calcutta from the East India Company's troops, commanded by John Holwell, the self-proclaimed governor of Bengal. Upon surrender, the nawab confined Holwell and many other Europeans overnight in the company's own prison cell, a tiny, dark room, roughly six meters (twenty feet) long and four meters (thirteen feet) wide, with two tiny windows, popularly known as "the Black Hole." Records of the incident from East India Company officials claim that 146 people were locked up in this minuscule cell, without adequate water and in extreme heat, and that only twenty-three survived. Although scholars such as J. H. Little have called these numbers into question, the Black Hole of Calcutta remains a powerful, macabre reminder of and sordid metaphor for the utter callousness of the nawab. Soon "Black Hole of Calcutta" entered the collective consciousness as a synonym for the most horrific of experiences. When a raging fire destroyed the Opéra comique building in Paris on May 25, 1887, a *New York Times* correspondent, noting that the seats, boxes, and balcony were gone, described the building as "an immense black hole."[2]

A black hole signifying a dark dungeon also appeared in literature well before the term took hold in physics. In the introduction to his horror story "The Premature Burial," published in the *Philadelphia Dollar Newspaper* in 1844, playing on the fear of being buried alive, Edgar Allan Poe mentions the suffocation and death of prisoners in the Black Hole of Calcutta.[3] So does one of the most imaginative writers of our time, Thomas Pynchon, who frequently cites the Black Hole of Calcutta in *Mason and Dixon,* variously as a musical drama and as a description of horror.

What is perhaps most surprising is just how well some of these descriptions fit an astrophysical object that had not yet been observed. In astronomy, a black hole is a physical location of no return. No one, however, is quite sure who first used the term to refer to these astrophysical miscreants. Tracking down its use, the science writer Marcia Bartusiak noted that although the physicist John Archibald Wheeler did not claim to originate the term, he most definitely popularized and legitimated this use.[4] He used the term during a lecture he gave in 1964 at a meeting of the American Association for the Advancement of Science. It stuck.

Today we know that black holes exist in the centers of most, if not all, galaxies. Our own galaxy, the Milky Way, harbors one such black hole four million times the mass of our sun. Farther afield, the glowing infalling gas sucked into the deep gravity of active and growing black holes renders them visible as quasars, some of the brightest beacons in the early universe. Quasars are visible from when the universe was barely 1 percent of its current age. From a nearly complete census of neighboring galaxies, we find that these often silent monsters lurk in galactic centers, revealing their presence only via their gravitational pull on the stars that orbit the innermost regions of galaxies. The immense gravity of these dormant (inactive) black holes is even apt to rip apart any stellar passerby that strays into their region of influence. Fortunately, our solar system is way too far from the center of the Milky Way for us to feel the presence of or be affected by its central black hole.

Astronomers now believe that black holes, despite their odd behaviors, are an inevitable consequence of the standard physics that describes the evolution of stars. The theory of stellar evolution predicts that stars born fifteen to twenty times more massive than our sun, after exhausting their fuel supply of hydrogen, will end their lives as black holes. Black holes may have exotic properties, but they are important constituents of the universe, playing a significant role in the assembly and evolution of galaxies.

Let us begin, then, the story of black holes when the term was still only a metaphor for oblivion and when their conception relied more on imagination than detection. This takes us to the ivory towers of Cambridge, England, in the 1700s. Cambridge and Oxford were closely tied to the Church of England then, and the typical student was from either the landed gentry or the clergy. This connection was attributable to a requirement that all graduates subscribe to the Thirty-Nine Articles (a statement of Anglican faith); therefore, it was not uncommon for the majority of graduates to end up going into the church.[5] It is with one such brilliant graduate, who imagined the existence of stars from which light might not be able to escape, that our tale begins.

In 1783, when an English country parson, John Michell, first proposed the idea of a "dark star," he could never have imagined that we would one day detect them. Michell, a polymath born in 1724, studied at Cambridge and later taught Hebrew, Greek, mathematics, and geology there. Although no portraits of him exist, a contemporary described him as "a little short man, of black complexion, and fat." A man of the cloth, he moved from Cambridge to a parish in Thornhill, near Leeds. Despite his religious commitments and duties, he was very much at the leading edge of science, and his reputation for originality was such that many of the active scientists of the day, the likes of Benjamin Franklin and Henry Cavendish, visited and maintained regular correspondence with him. They had much to discuss; Michell's scientific contributions include describing the strength of magnetic fields and developing a theory for how earthquakes propagate through faults on the earth's surface. This seismological research led to his election to the Royal Society in 1760. Despite his achievements as a natural philosopher, he remains less known than his contemporaries, because he did not promote his ideas sufficiently.[6]

He thought of light, as Newton had postulated, as particles, or "corpuscles." Michell proposed that massive stars could gravitationally attract and slow down light particles, much as they attract other astronomical passersby, such as comets. Since the more massive the star, the stronger its gravitational pull, some extremely massive stars, he noted, might stop light entirely. In a letter to Henry Cavendish dated November 27, 1783, Michell anticipated that such "dark stars" would be observable only by the impact they had on bodies revolving around them. He subsequently published his idea, a Newtonian precursor for the black hole, in a paper in the *Philosophical Transactions of the Royal Society of London*. Michell was not alone; thirteen years later, in *Exposition du système du monde* (1796), the French mathematician Pierre-Simon Laplace suggested a similar concept, concluding, "It is possible that the greatest luminous bodies in the universe are on this account invisible." Yet when Newton's corpuscular theory of light later lost favor, so too did the idea of dark stars. In fact, later editions of Laplace's book omit the discussion of dark stars entirely.[7]

It took 150 years and Einstein's theory of general relativity to resurrect the idea of such an astronomical object. General relativity grew from a simpler idea. In 1905 Einstein postulated the theory of *special* relativity. His conclusion: nothing can travel faster than the speed of light. This universal speed limit has profound implications. For one, it sets the maximum speed at which any matter or information can travel. It also establishes the equivalence between mass and energy, summarized in the famous equation $E = mc^2$. It was Einstein's 1915 theory of *general* relativity, however—which profoundly transformed our understanding of mass, gravity, and space—that brought black holes back into the fold.[8] The mathematical rigor of general relativity enabled an entirely new way of visualizing reality. As we saw in the previous chapter, it led to the creation of a new model of the universe, the first major revision since Newton. But much to Einstein's dismay, his theory of general relativity also allows for the existence of black holes.

At the risk of portraying Einstein as an indefatigable curmudgeon, we must note that, much as he resisted the concept of an expanding universe, he also hated the idea of black holes. Part of physicists' unadulterated admiration for Einstein is due to his development of the entire theory of general relativity ex nihilo, not to provide an explanation for any observed phenomenon but as a self-sufficient, fundamentally new theory of gravity. This is as pure as theory gets in physics. General relativity is an homage to the power of speculative thought and the possibilities of deep mathematical understanding. The theory offers profound insights into the nature of gravity, the mysterious attractive force that holds the solar system and the entire universe together. An impulse to find simplicity and unity in nature motivated Einstein throughout his scientific career. This same philosophical inclination occasionally proved a hindrance to his accepting complexities when they arose—even in his own theories and work. Such was the case for black holes.

Einstein's theory is mathematically elegant and does not hinge on any scientific observations, but it did make several testable predictions. Initially well ahead of any observational revelations or applications, after its veracity was tested, the theory of general relativity became a somewhat sterile topic, out of the mainstream of scientific investigation. Einstein's theory has significant implications for astronomy, but its connection to existing physical objects seemed tenuous at best in the early 1900s. The need for general relativity to explain compact objects did not emerge for some time, although it was successfully used to describe the overall properties of the universe within a decade of its invention. Because the observational effects of this theory are small unless the objects studied have extreme gravity, it was not until astronomers discovered exotica such as neutron stars, pulsars, and quasars that its full explanatory power was revealed. So when astronomers detected these heavyweights in the universe in the 1960s, Einstein's theory was already firmly in place to describe their properties.

Today some of the most convincing evidence for the existence of black holes comes from the spiral galaxy NGC 4258. Inside NGC 4258 is a black hole about forty million times more massive than our sun. To give a sense of scale, mapping the innermost regions of this galaxy using radio waves, astronomers have uncovered a disk that appears to be a reservoir for the gas that is likely swirling down into the black hole, so wide that it would take light a year to cross it (if it weren't captured). Black holes of comparable mass lurk in the centers of many nearby galaxies, where they affect the motion of stars around them. Supermassive black holes with about a billion times the mass of the sun have been inferred to exist in the centers of the brightest galaxies.[9]

To appreciate why and how black holes have the properties that they do, we need to understand gravity as explicated by Einstein. Gravity is a universal force, and although as forces go it is not very mighty, nothing is exempt from its grip, be it stars, planets, or galaxies. Newton was the first to understand gravity's attractive nature and that it is responsible not only for holding us on earth but also for holding the planets in their orbits. The more massive and compact a body, the stronger its gravitational pull. Because black holes are massive and extremely compact, they exert the strongest gravitational pull in the universe. As you might remember from high school physics, we refer to the speed required to break free from the gravitational pull of an object as its escape velocity. For example, for a rocket to escape earth's gravity, its velocity needs to be an impressive forty thousand kilometers (more than twenty-four thousand miles) per hour, which is what rocket boosters provide to launch satellites from Cape Canaveral in Florida, the Baikonur Cosmodrome in Kazakhstan, and Sriharikota in India. By comparison, the escape speed from the sun, about 330,000 times the mass of the earth, is about a hundred times larger, around four million kilometers (about 2.5 million miles) per hour—still 250 times smaller than the speed of light. What happens when the escape speed from a body equals or exceeds the speed of light? This question puzzled Michell as he sought to understand the propagation of light from stars. The answer: black holes. Not even reflected light divulges the presence of black holes. Nor are they just stars cloaked because of extreme light bending. Their strong gravitational pull transforms the shape of space and strangely disrupts the flow of time in their proximity. This is why, to understand black holes, we need to think like Einstein.

His seminal 1905 paper in *Annalen der Physik* contains a remarkable piece of insight.[10] Here, Einstein provided a profound new theory that completely reconceptualized our understanding of the relationship between mass, gravity, and space. Newton saw gravity as an attractive force, transmitted instantaneously between any objects with mass, but special relativity's universal speed limit makes the instantaneous impossible. In contrast, according to general relativity, objects with mass generate a gravitational field, which in turn is related to the shape of space. In this picture, gravity is better understood not as a pull but as a distortion of space itself, altering how other bodies are constrained to move in response around that mass. Central to this idea is the notion of space-time. The entire universe and all its contents—galaxies, stars, and planets—inhabit space-time. Space-time can be thought of as a sheet on which masses produce divots that impact motions as well as the flow of time. Another way to visualize space-time is to imagine it as a landscape—like a topographic map—with valleys generated at the locations of masses.

The leap that Einstein took from Newtonian gravity to his theory of general relativity is one of the rare examples of inductive reasoning in science. Although not inspired by observations, Einstein's

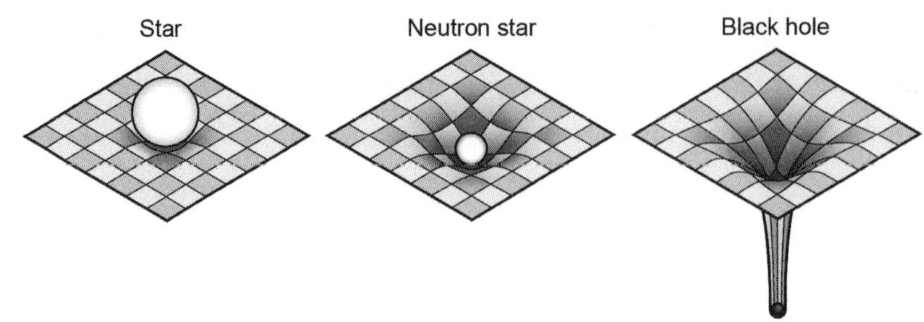

Distortions in space-time produced by objects of increasing mass and compactness, as predicted by Einstein's theory of general relativity.

pure theory made concrete and testable predictions that helped evaluate it and establish its validity. This might seem unfamiliar compared to the usual relationship between observations and theory, wherein theories are formulated to explain observational facts via deductive reasoning.

General relativity predicts gravitational lensing—as mass distorts space, tracking the distortion, light bends. When the earth and the sun align during a solar eclipse, the divot caused by their masses is deeper, which makes the bending of light observable. The astronomer Erwin Finlay-Freundlich proposed the first experiment to verify the theory of general relativity, by observing the bending of starlight during a total solar eclipse. Presenting his proposal to make this astronomical observation, he informed Einstein that the next observable total eclipse would be visible from Crimea in the summer of 1914. Einstein even helped raise funds for the expedition, which set off early that summer, but the outbreak of World War I ended the quest. Much to the shock of the pacifist Einstein, the Russian army in Odessa arrested and imprisoned the entourage. This delayed verification until 1919, when Arthur Eddington led one of two British expeditions during a solar eclipse to measure the deviation in the path of light rays that passed close to the sun. Eddington photographed and measured the positions of several stars near

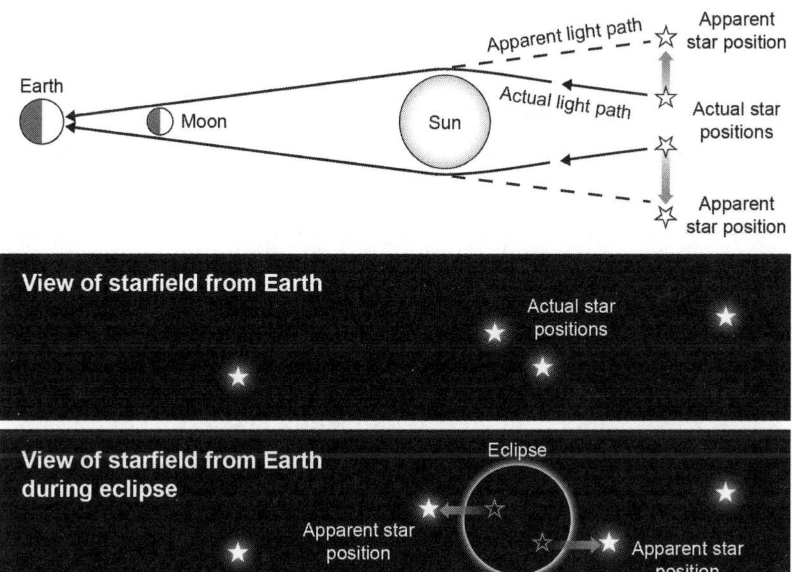

Schematic showing how gravity can bend light, as measured by Eddington during his 1919 solar eclipse expedition. The real and apparent positions of stars were observed to exactly match the predictions from Einstein's theory of general relativity.

the sun during the eclipse. Light from these distant stars traversed space-time that was distorted by the gravity of the sun. These stars appeared displaced from their positions measured six months before the eclipse. The presence of the sun had indeed deflected the light by the amounts that Einstein's theory exactly predicted. Eddington's November 6, 1919, announcement of the discovery of light bending to a joint meeting of the Royal Society and the Royal Astronomical Society made Einstein instantly famous. Yet confirmation of this theory did more than make Einstein an icon; it opened up investigation of further potential applications.[11]

It is worth reiterating that general relativity was proposed well ahead of any anticipated uses, although today it is necessary and in fact indispensible both for theory and in applications including the global positioning system (GPS) in your phone and the calculations that helped to land a rover on the surface of Mars.

Although Newton's law of gravity offers an accurate description of how objects fall to the ground, we now know his description is not complete or comprehensive. Newton's law fails, for example, to describe the motion of particles on the smallest atomic scales or on the largest scales in the cosmos. To understand these, we need to defer to Einstein's interpretation of gravity. He could not have anticipated our current reliance on his theory of general relativity to get our bearings. GPS technology is built entirely on Einsteinian gravity. Determining accurate positions and navigating on earth today is made possible by a fleet of twenty-four satellites that orbit the planet, each carrying one of the most accurate clocks available—atomic clocks. A GPS receiver in a navigator in your car on earth picks up radio signals from the satellite that is closest by and trilaterates with signals from four other satellites. This enables the computation of your current position to an accuracy of one meter (three feet) or less. This is no mean task, as it requires taking into account corrections predicted by both of Einstein's theories—the special and the general theory of relativity. According to the theory of special relativity, moving clocks tick more slowly than stationary clocks. Therefore, clocks on satellites tick more slowly than clocks on the ground. Einstein's theory of general relativity meanwhile predicts that clocks circling above the earth immersed in its gravitational field should tick faster. This is due to the fact that gravity curves space *and* modifies the flow of time. The combination of these two competing effects causes an orbiting clock to be ahead very slightly—by about forty microseconds per day. Tiny as this correction might seem, it is of great consequence for positional accuracy—without this correction you might end up in New Jersey instead of Manhattan—certainly a significant difference! Einstein's theories do not invalidate Newton's conception of gravity—they both have their domains of validity. Each theory has its place and offers adequate and accurate descriptions, in certain regimes. As Einstein once said, "The most beautiful fate of a physical theory is to point the way to the establishment of a more inclusive theory, in which it lives as a limiting case." Newton showed gravity's universality; Einstein's theory explains why this is so in terms of the curvature of space-time.[12] For example, within the solar system the deviations from Newton's theory predicted by general relativity are tiny, about one part in a million or so.

As in the discovery of the map of our solar system, irregular planet motions again come into play. Einstein postulated that one other testable, observational consequence of his theory would be the precession of the orbit of the planet Mercury. The motion of planets in the solar system drags along the bumpy fabric of space-time. Mercury, because of its closeness to the sun, is impacted more acutely than planets farther away by the divot of space-time generated by the gravity of the sun. This causes tiny but measurable anomalies in its orbit, which were found to be in agreement with Einstein's prediction. More recently, precision experiments with space probes have verified this key prediction of general relativity—the modulation of the orbit of Mercury—to an extremely high degree of accuracy.

Einstein did not believe, however, that general relativity's field equations describing gravity would admit any simple solutions. But in 1915, the German physicist Karl Schwarzschild discovered an exact solution for a special case of space-time surrounding a tiny yet massive object. This solution describes a distortion or modification of the shape of space, the divot around a point mass: a black hole. Since then, physicists have found several other exact solutions to Einstein's field equations. As we saw in the previous chapter, Alexander Friedmann and Georges Lemaître derived another solution, which shows that space-time, meaning the entire universe, expands. And as recently as 1963, the relativist Roy Kerr found a solution that describes a rotating black hole.[13] Schwarzschild's black hole solution, although mathematically exact (and not an approximation), was rather peculiar and non-intuitive to physicists. What is odd about this solution is that the black hole encases a singularity, a point where the laws of physics as we know them break down and are no longer valid. Another bizarre feature of Schwarzschild's solution is the existence of a boundary between a black hole's visibility and its invisibility—what is now called the event horizon or Schwarzschild radius. This boundary demarcates a point of no return. Any object or light signal that passes through the event horizon of a black hole is lost forever, unrecoverable. Furthermore, this boundary occurs precisely at a radius proportional to three times the mass of the black hole. Thus, the bigger the black hole, the bigger the radius. Physicists regarded Schwarzschild's solution, including its properties of an event horizon and a concealed singularity, as a mathematical curiosity—most definitely not the representation of any real objects that might exist.

One of the key reasons that black holes were unpalatable to physicists is the deeply problematic nature of the singularity. Singularities are challenging because their existence tests the limits of theories and denotes a realm where our intuitions also cease to be useful. Physicists still have to live with the uncomfortable notion of the singularity, since they understand that it is inevitable given the steep incline of the space-time divot around the black hole. This is a limitation that cannot be sidestepped but one that physicists hope to overcome by formulating a unified theory that synthesizes the physics of the smallest scales—quantum mechanics—with gravity. Several generations of physicists, including Einstein and Eddington, have dreamed of such a final theory, the so-called theory of everything; however, it has remained elusive. Yet the understanding

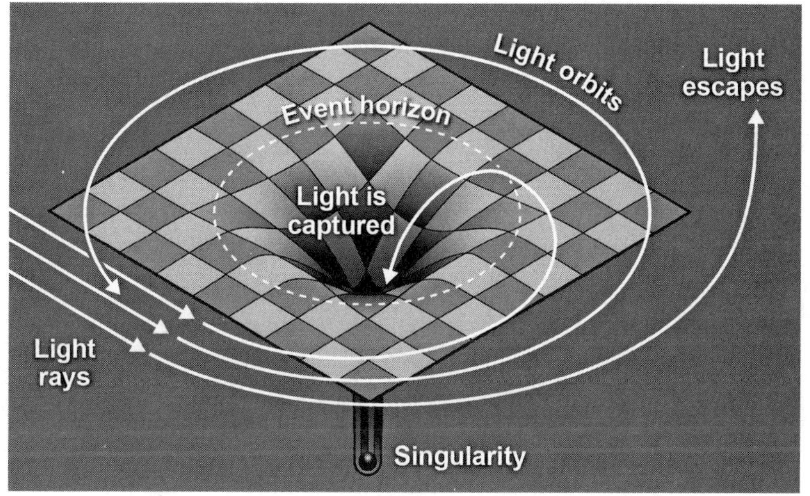

Schematic of a black hole with its event horizon, showing the fate of approaching light rays. The bending and trapping of light rays depends on where they impinge relative to the event horizon.

that the singularity lay inside and not at the horizon was a major breakthrough, because it provided clues to how black holes actually form, for example at the end of a star's collapse.

The path from dying star to black hole requires further explanation. Imagine a typical, so-called main sequence star, such as our sun. Its interior is extremely hot, much hotter than its glowing surface. The interior has a temperature of fifteen million kelvin (150 million degrees Celsius or 270 million degrees Fahrenheit). It's so hot that particles smaller than the atom—electrons and nuclei—are bouncing around and running into one another constantly. These collisions generate pressure inside the star. The pressure from the star's innards counterbalances the effect of gravity and thus halts the star's natural collapse.

Yet in the case of stars, such equilibrium, due to the delicate balancing of forces, cannot last indefinitely. The presence of an energy source at the center of the sun, the fusion reactor that converts

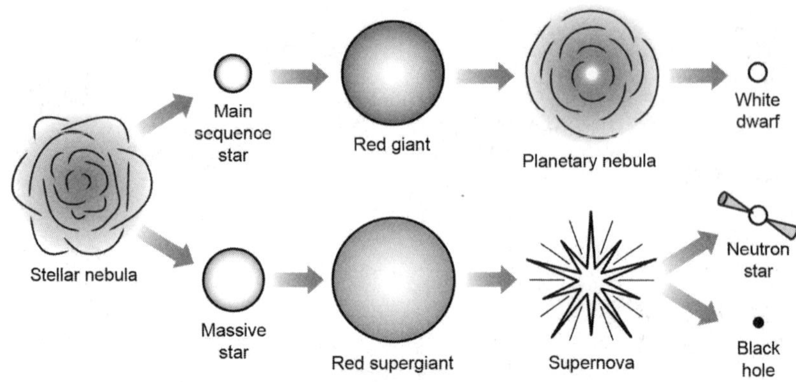

Schematic of the stages of stellar evolution. Depending on the initial mass of the star, stellar evolution theory predicts that the end point will be a white dwarf, a neutron star, or a black hole.

hydrogen into helium, keeps the forces balanced to start with. But once fusion consumes all the hydrogen in the core, gravity will win the race and pull the core in farther. At this point, the heavier chemical elements may start to fuse, but eventually the star will run out of all its nuclear fuel and begin to cool down. As is expected to happen in about five billion years, our sun will cool down and become a white dwarf as gravity wins the balancing game. The fate of stars more massive than our sun, though, is more exotic—they continue to collapse and contract much farther, becoming either neutron stars or black holes.

The push that resurrected interest in black holes came from the work of the theoretical astrophysicist Subrahmanyan Chandrasekhar, known as Chandra, who worked out his ideas about them on his maiden voyage from Madras, India, to Cambridge, England, to study at Trinity College in 1930. He figured out that under some special circumstances, at the end of stellar evolution, when all the fuel powering nuclear fusion has run out, a star might end up producing an extremely compact object. Chandra's calculation shows explicitly

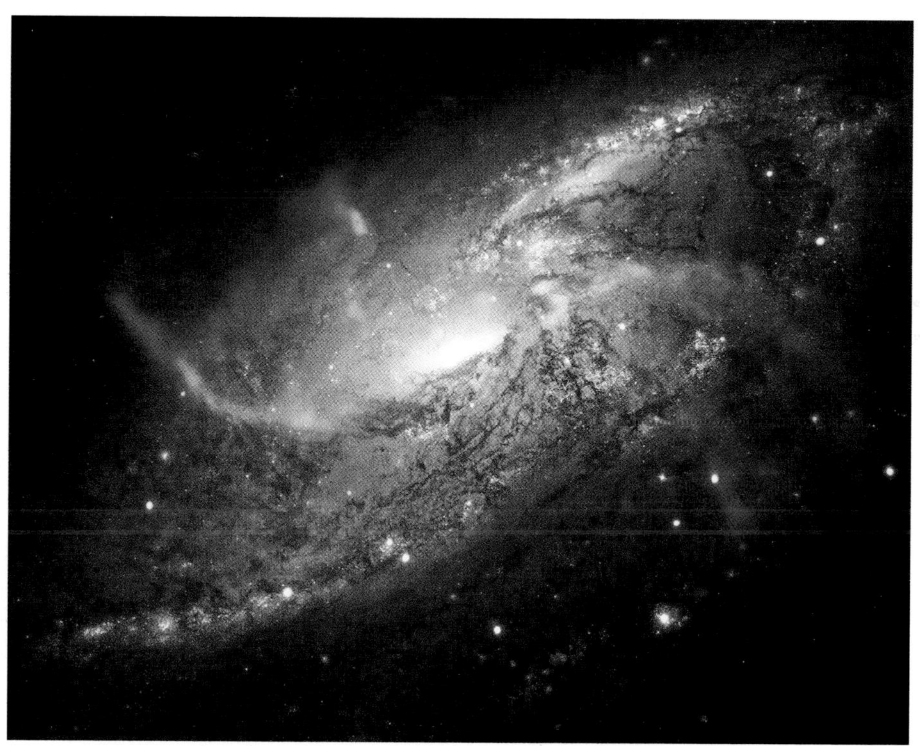

Multiwavelength composite view of the galaxy NGC 4258, which hosts a black hole four billion times the mass of our sun. Created from X-ray data from the Chandra X-ray telescope, an optical image taken by the Hubble Space Telescope, an infrared image from the Spitzer Space Telescope, and a radio image from the Karl G. Jansky Very Large Array.
Courtesy of NASA/CXC/Caltech/P.Ogle et al. (X ray); NASA/STScI and R. Gendler (optical); NASA/JPL-Caltech (infrared); and NSF/NRAO/VLA (radio).

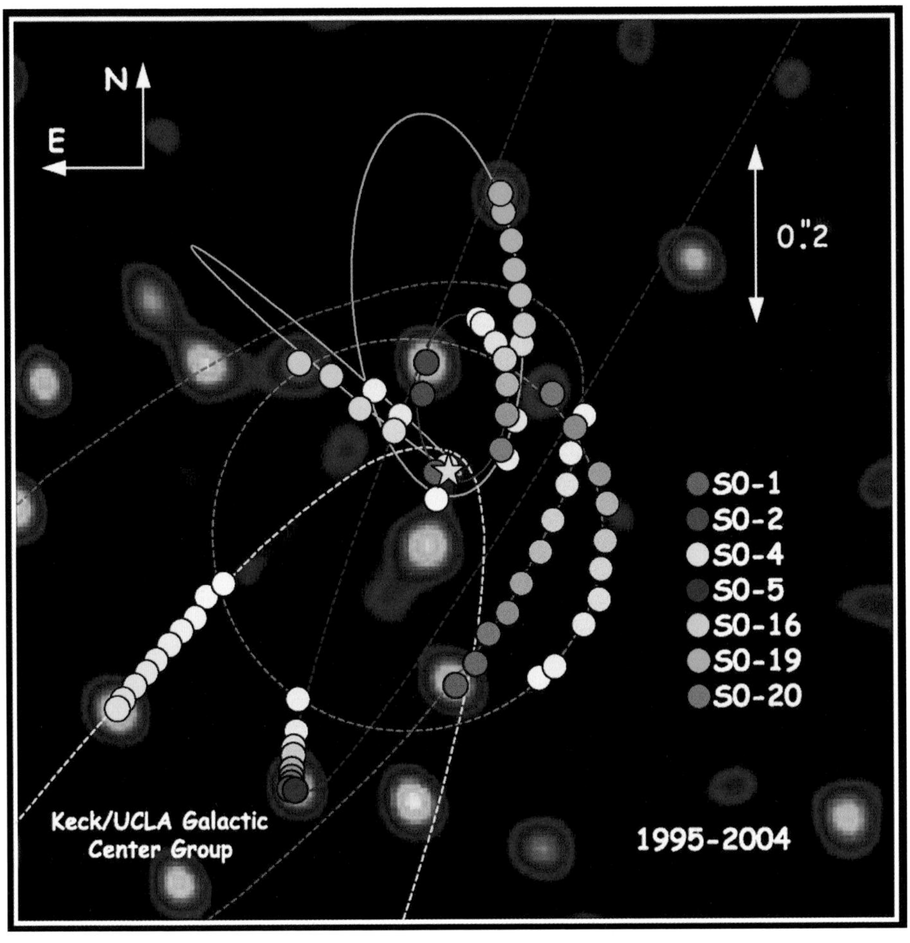

The orbits of stars in the center of our galaxy. Created by Professor Andrea Ghez and her research team at UCLA with data sets obtained with the W. M. Keck telescopes.
Courtesy of the UCLA Galactic Center Group, W. M. Keck Observatory Laser Team.

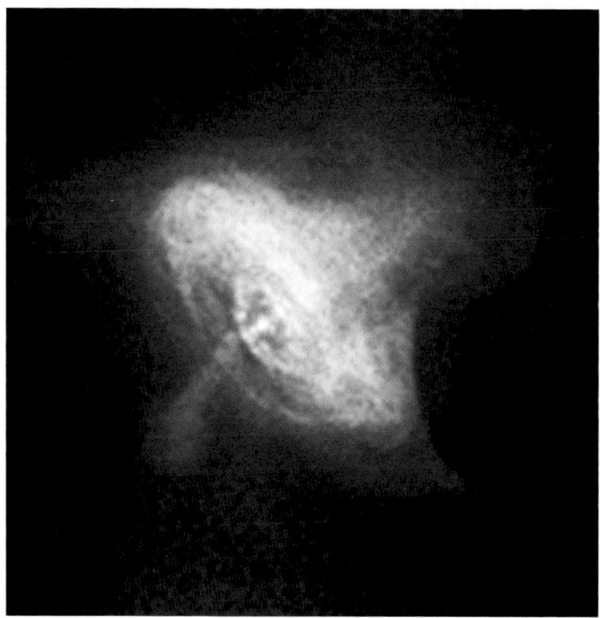

The Crab Nebula: top, an image from the Hubble Space Telescope; and bottom, an X-ray image from the Chandra telescope. Courtesy of NASA, ESA, J. Hester and A. Loll of Arizona State University (Hubble); and NASA/CXC/SAO (Chandra).

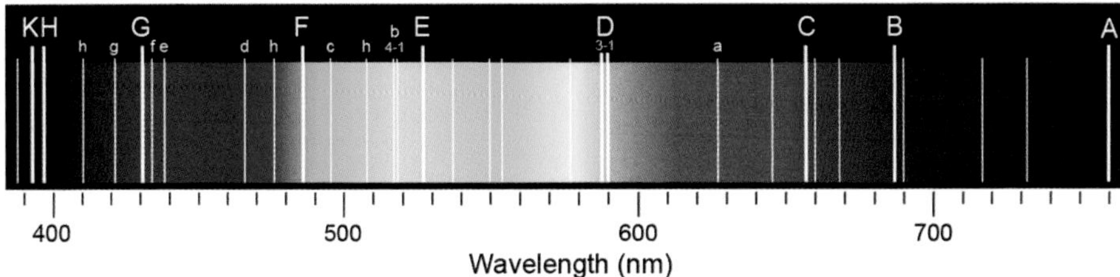

Fraunhofer lines in the solar spectrum. These absorption lines indicate the presence of chemical elements in the sun and represent their unique spectral signatures.

Opposite: Images of two gravitational lenses from the Hubble Space Telescope. *Top to bottom:* The massive lensing cluster Abell 2218, replete with numerous lensed arcs; the recent Frontier Fields project deep image of the lensing cluster MACS0416, taken in 2014; and the reconstructed dark matter map of MACS0416, in blue overplotted on the Hubble data. Abell 2218 image courtesy of NASA, ESA, Richard Ellis (Caltech), and Jean-Paul Kneib (Observatoire Midi-Pyrénées, France); MACS0416 Frontier Fields image courtesy of ESA/Hubble, NASA, HST Frontier Fields and J. Lotz, M. Mountain, A. Koekemoer, and the HFF Team (STScI); MACS0416 dark matter map courtesy of Mathilde Jauzac (Durham University, United Kingdom, and Astrophysics and Cosmology Research Unit, South Africa) and Jean-Paul Kneib (École Polytechnique Fédérale de Lausanne, Switzerland).

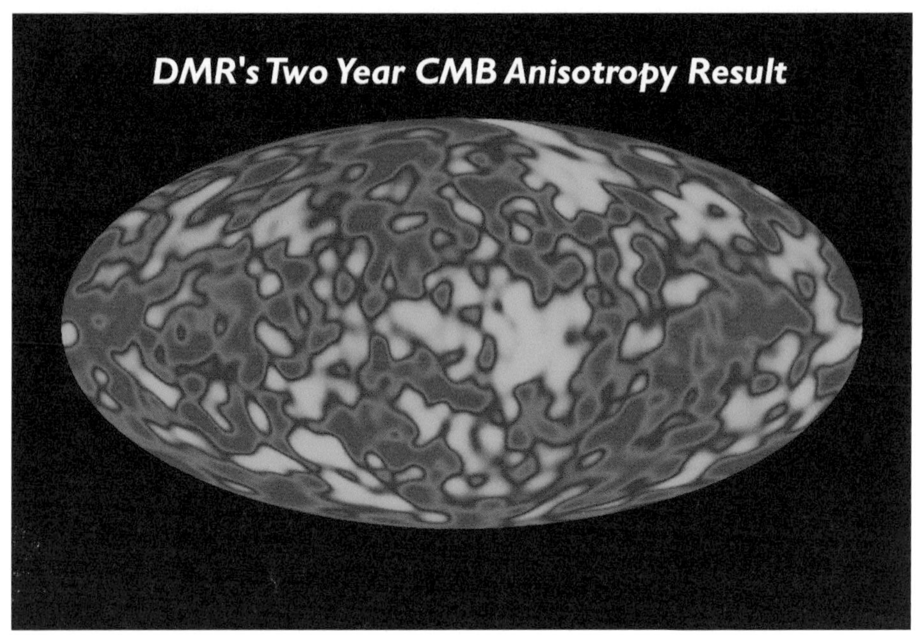

The map of the CMBR as detected by the COBE satellite—the map of temperature fluctuations measured in the microwave sky, pockmarked with hot spots (pink pixels) and cold spots (blue pixels). Courtesy of NASA/GSFC.

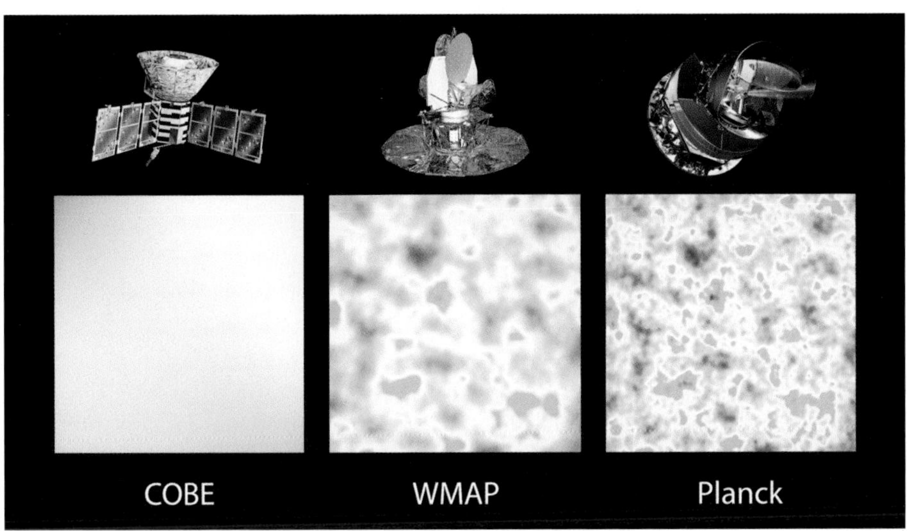

Comparison of the resolution of CMBR fluctuations detected by the COBE, WMAP, and Planck satellites. Courtesy of NASA-GSFC.

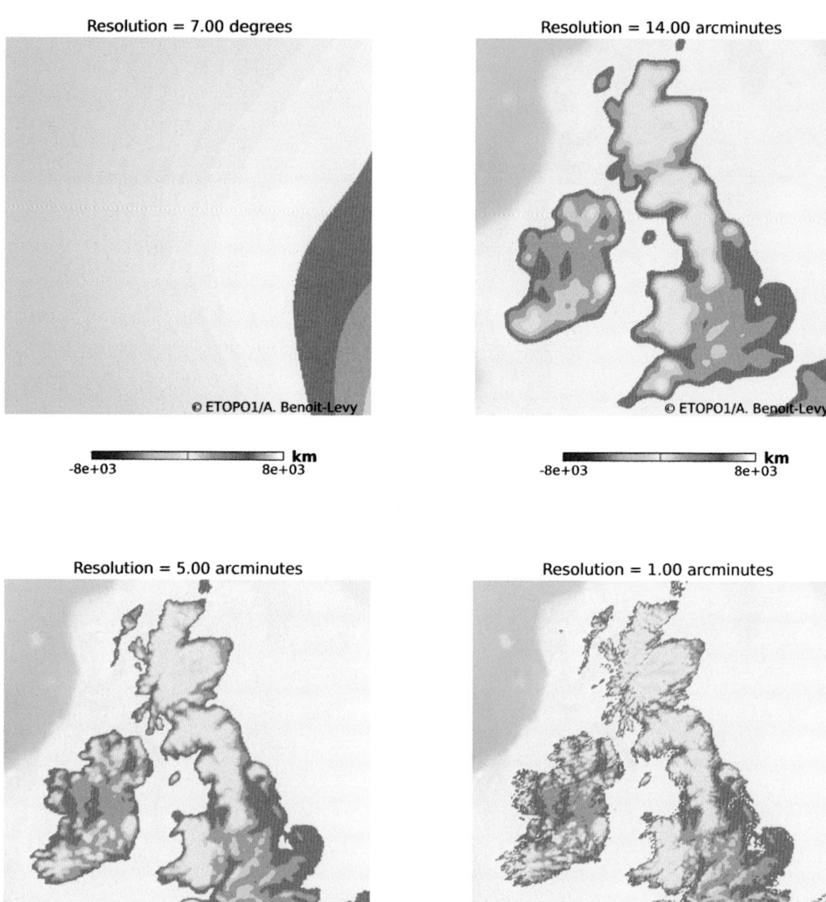

Visualization of the coastline of the United Kingdom to demonstrate the difference in resolution of the microwave background fluctuation maps created by the COBE, WMAP, and Planck satellites.
Courtesy of Aurélien Benoit-Lévy, University College London, using data from NOAA (EPTOPO1 map of U.K. coastline) and the National Geophysical Data Center.

that some stars will get to this uncomfortable end point, becoming an infinitely small and infinitely dense entity—the singularity, which we now call a black hole. Chandra married two fundamental theories—the general theory of relativity and quantum mechanics—to derive the limiting mass of a star that would implode and transmute into a black hole. There was, however, strong opposition to his entire model of stellar death, not just to the conclusion that black holes might form. Later work expanded Chandra's model and found that stars that were initially 1.4 to 3 times more mass than the sun would end up as neutron stars while those that are ten to twenty-five times more massive than the sun are the ones that would end up as black holes.

Ironically, the most virulent intellectual opposition to Chandra's ideas came from his colleague Eddington, an early verifier and supporter of Einstein's theory of general relativity. One would imagine that someone who had been so open to accepting the radical idea of general relativity and so instrumental in proving it would have been more enthusiastic about Chandra's conclusions. But Chandra's calculation posed a tremendous conflict of interest for Eddington, who had been developing his own theory, also a union of Einstein's general relativity and quantum mechanics, to calculate what happens when stars collapse under their own gravity. He saw his bold new theory as the culmination of his life's work, his legacy that would combine the largest and smallest scales in the universe—the subatomic world and the cosmos. His conception did not include black holes. Eddington did think that a small, dense star would distort space around it so much that it would prevent light from escaping, but he surmised that this strange object and deep divot would simply disappear and, in his words, be "nowhere." The notion of a singularity was so abominable that even Einstein thought, incorrectly, that black holes would not form, that there had to be some mechanism that would stabilize a collapsing star before it reached this point of no return. Both he and Eddington firmly believed that nature simply would not permit such a perverse fate for stars.[14] They regarded the notion of a black hole as an imperfection that needed to be excised from the theory and not as an inevitable and testable consequence.

In a famous showdown, Eddington unexpectedly tore into Chandra at a meeting of the Royal Astronomical Society in 1935. He had been aware of Chandra's calculation because they were both fellows at Trinity College, Cambridge, and talked science frequently. However, he had not mounted his objections during those discussions. Rather, Eddington, who was intellectually powerful as the director of the University Observatories at Cambridge, chose to voice his criticism at a public meeting, one of the monthly get-togethers of all the prominent astronomers in Britain. On that fateful day, January 11, 1935, Chandra delivered his paper showing a graph of the fate of stars above a certain mass, which would leave behind peculiar and bizarre corpses, black holes. After presenting this radical new idea, he was hopeful that Eddington, who spoke afterward, would support and endorse his claim. After all, they had discussed Chandra's theory prior to this meeting. Besides, Eddington had been one of the examiners for his PhD oral exam and had previously been encouraging and enthusiastic. To Chandra's utter shock, Eddington, with his usual eloquence and authority, ripped into the idea, declaring it nothing more than slippery mathematics that had nothing to do with reality. His objections were baseless, yet given the formal nature of the meeting, Chandra did not have the opportunity to respond to the criticism. Although many other eminent physicists of the day, including Chandra's thesis adviser Ralph H. Fowler, Leon Rosenfeld, Wolfgang Pauli, Paul Dirac, and Bill McCrea, were in the audience and supported Chandra intellectually, they did not dare oppose Eddington in such a public forum. It was a moment of intense betrayal for Chandra, who felt humiliated and unsupported by his thesis adviser Fowler and the larger British physics community. Three of the most eminent physicists of the time—Dirac, Rudolf Peierls, and Maurice Pryce—though, wrote an important paper in 1942 that supported his view.[15]

Arthur I. Miller presents a poignant and detailed account of this dramatic story in his book *Empire of the Stars*. He describes this episode, its fallout, and the tale of Chandra's struggle against the establishment, epitomized by Eddington, as a case study of how science works and how clashes of ideas can play out. Kameshwar Wali's detailed biography of Chandra, which spans his life from childhood through his scientific career, also discusses this episode and its lasting impact on Chandra. Meanwhile, for us this is also important as an instance when complicated personal stakes intervened in science, superseding intellectual ones. Eddington's objections stemmed from his personal distaste for singularities and from the threat that Chandra's work posed to his own model, which he viewed as his bequest, neither of which has intellectual standing. Despite the far-reaching consequences of this controversy for Chandra, he and Eddington remained personally cordial, albeit with awkwardness throughout it all. Miller minces no words in describing Eddington's behavior as sheer, mean-spirited duplicity. He also uncovers from Chandra's letters that Chandra resorted to underhanded dealings himself to publish his papers refuting Eddington's theory. He set about appeasing one of Eddington's archrivals, the physicist Sir James Jeans, to garner support and get his papers favorably refereed.[16] Eddington remained wedded to his theory and never relinquished his position. Both men were blinded by their emotional attachment to their respective ideas.

As always in science, data and evidence in the end determined who was right. The physicist Max Planck's lament, in response to resistance to his own scientific ideas, that "a new scientific truth does not triumph by convincing its opponents and making them see the light, but rather because the opponents eventually die, and a new generation grows up that is familiar with it," was borne out in this case.[17]

Ultimately, it was a strange twist of fate that established the importance and validity of Chandra's ideas. It was only during the arms race in World War II, when computer calculations showed that the core of a hydrogen bomb essentially resembles an exploding star, that physicists realized the relevance of his calculations. Chandra prevailed, winning the Nobel Prize in 1983 after astronomers validated his theory with the discovery of neutron stars. They also discovered pulsars, the stellar equivalent of lighthouses, in 1967. Within two years of this discovery, it was shown that pulsars are rapidly rotating neutron stars containing as much matter as our sun but packed so tightly that they have the density of the nucleus of an atom. This packing, it turns out, is close to the critical density, right at the margin at which gravity would overwhelm the star and cause it to collapse into a black hole. This discovery renewed the astronomical community's interest in the study of compact objects that form as a result of gravitational collapse. That neutron stars were an astrophysical reality meant that their cousins, black holes, likely existed as well.

Because a black hole does not emit any light, we cannot observe it directly. But as even Michell suggested, black holes can "reveal" themselves by their action on surrounding matter. Therefore, when a black hole orbits another star, its immense gravitational pull rips gas from its neighbor. Gas funneling into the black hole heats rapidly, making it glow and emit radiation in the form of X rays. We have observed such systems of a star and its black hole companion. These observed X-ray binaries propelled black holes into the realm of respectability.

By extension, when astronomers discovered quasars, it became clear that these were powered by more-gigantic, supermassive black holes, glowing while feeding on gas in their vicinity too. Quasars are the brightest objects in the universe. We have detected large populations of these supermassive black holes, and our current understanding is that every galaxy likely underwent one of these bright, quasar phases in its history—a time when the black hole was actively feeding on gas before exhausting the available supply.

The other indirect way that researchers have identified black holes in the centers of nearby galaxies and our own is by quantifying their gravitational influence on stars that orbit the galaxy's center, which allows for the calculation of their mass. Astronomers have now mapped the orbits of stars immediately around the black hole at the center of the Milky Way, and their paths clearly reveal the presence of a monster lurking in the middle. We can procure this level of detail only for our galaxy, because distance precludes the study of the inner regions of even nearby galaxies.

Every year brings new evidence. In early 2014, a gas cloud made a close passage by the black hole at the center of our galaxy. During this encounter we expected the black hole to disrupt and swallow the cloud. Such an event would produce a dramatic, visible flare that would illuminate a black hole feeding frenzy—but we did not witness any such drama; the cloud simply slunk by. This was a unique opportunity for scientists to witness a direct test of the strong gravity of a black hole. As it turned out, the gravity of the black hole had less of an influence on the cloud than was theorized. Therefore, the data of this encounter has called the nature of this cloud further into question. Although this quest for direct access was unsuccessful, astronomers are pursuing another avenue to garner more information about the black hole indirectly. This new and exciting project, currently under way, aims to map the black hole at the center of our galaxy with a new instrument—the Event Horizon Telescope (EHT). Due to their extreme warping of space-time, black holes exert a peculiar influence on time as well as the propagation of light in their vicinity. The dramatic deformation of the fabric of spacetime around a black hole causes light that grazes nearby to reflect bizarrely, producing a unique kind of shadow that skirts the event horizon. The EHT uses radio telescopes in Mexico, Chile, and Germany to map the shadow around the black hole at the center of our galaxy in radio wavelengths. These widely dispersed telescopes are networked to act as one single telescope with a collecting area almost the size of the earth's face. This clever trick will provide the sharpest possible view of a black hole's shadow. Mapping the shape of this shadow, including any asymmetries or elongations, will reveal the properties of the black hole, for instance if it is spinning or not. The spin of a black hole is one of its key defining properties, along with its mass. The EHT is the most novel, most recent indirect approach to mapping black holes. But our attempts at indirect mapping have a long and rich history.

Thus far we owe the most convincing evidence for black holes to the discovery and understanding of X rays. It all started in 1895, with the discovery of X-ray radiation by Wilhelm Röntgen, a well-respected experimental physicist and the head of the physics institute at the University of Würzburg in Bavaria. He was interested in figuring out if the recently discovered, negatively charged cathode rays (beams of electrons) were particles or waves. We now know they are both, thanks to quantum mechanics, but Röntgen lived in a prequantum world. Working late on a Friday evening in his laboratory, he examined cathode rays hitting a fluorescent screen. Despite having completely darkened the room and covered the tube to block all light, he noticed a flickering spot on the screen, which was a few feet away. He checked the blackout curtains on the windows, but they were tightly closed. No light strayed into the lab from outside. Holding a lead sheet in the path of the beam, he saw a clear image of the bones in his hand beside the shadow of the sheet. Experimenting further on the same evening, of November 8, 1895, Röntgen

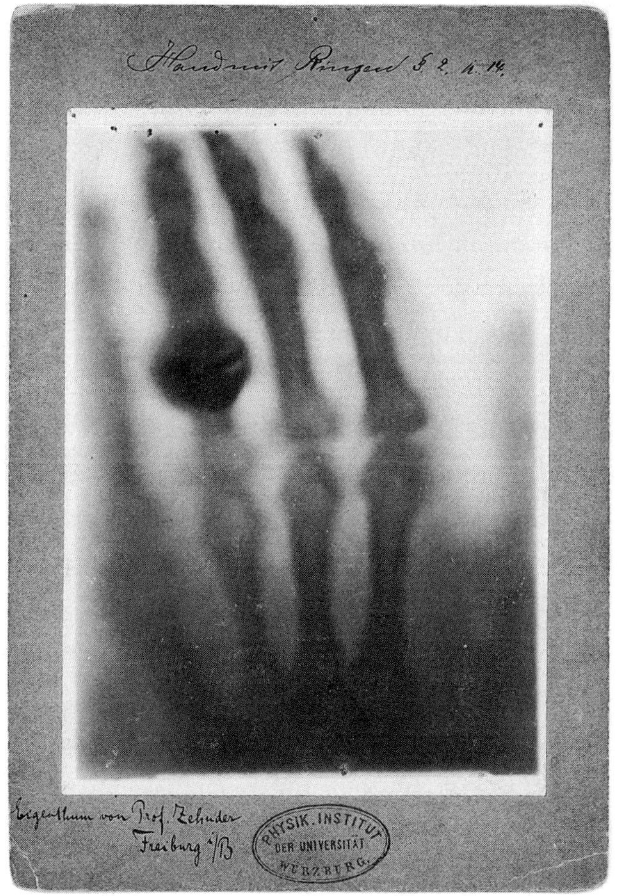

X-ray image of Anna Bertha Röntgen's hand—one of the first produced. Courtesy of the German Röntgen Museum, Remscheid.

figured out that this radiation, which he dubbed X rays, was emitted when the cathode rays (electrons) hit the wall of their glass tube. He noted that they were extremely powerful, as they penetrated human skin and directly imaged the bones underneath. Excited, he recorded the first X ray, of the left hand of his wife, Anna Bertha Röntgen (née Ludwig), which reveals the bone structure and the shadow of her wedding ring.

After further intensive study, Röntgen finally published his results early in 1896, and his discovery caused a sensation around the globe. William Thomson, 1st Baron Kelvin, one of the leading physicists of the time, initially thought that Röntgen's paper was a hoax, changing his mind only after worldwide replications of the experiment. Röntgen was awarded the inaugural Nobel Prize in Physics, for the discovery of X rays in 1901. This key discovery eventually led to the unveiling of black holes in the universe.

X rays are a form of extremely energetic electromagnetic radiation. The electromagnetic spectrum consists of radiation with small to very large wavelengths. X rays are the most energetic part of the spectrum, with the shortest wavelengths (in the range of 0.1 to 1 nanometer[18]); by comparison, visible light has wavelengths of 390 to 700 nanometers. In contrast, radio waves have the largest wavelengths (1 millimeter to 100 kilometers, or 62 miles). Because the human retina is not sensitive to X rays, we can see them only by using special detectors.

The discovery of neutron stars and pulsars suggested that the theory of stellar evolution's prediction of stellar death was valid and that black holes were an inevitable outcome under some circumstances. This eventually spurred the observational hunt for black holes. Nature, it turns out, provided a critical clue early on: the fiery, dying gasp of a star in 1054 CE, a supernova explosion observed and recorded assiduously by the Chinese. The court astronomer Yang Wei-Te reported the appearance of this bright new guest star in the constellation of Taurus to the emperor. We still see the afterglow of that explosion today, as the Crab Nebula, whose original star ended up as a pulsar, surrounded by a glowing, expanding debris shell.

Stars less massive than our sun leave behind a white dwarf as a corpse. Stars weighing more than the sun are too massive to become white dwarfs after burning up all their nuclear fuel. Such stars have a core that implodes, as a result of which their outer layer is expelled as a supernova. This stellar detritus contains all the chemical elements that make us. All the calcium in our bones, for instance, was once synthesized inside the cores of stars and spat out violently during such supernovae explosions on their demise. The theory of the birth, evolution, and death of stars predicts that what is left behind after the implosion of a massive star is either a neutron star or a black hole. When the Cambridge graduate student Jocelyn Bell and her thesis adviser, Anthony Hewish, discovered pulsars in 1968, this result leaped from the theoretical to the observed. While surveying the sky with a new radio telescope commissioned at the Mullard Radio Astronomy Observatory in the outskirts of Cambridge, Bell and Hewish found a source that was emitting regular pulses every 1.3 seconds. Their search turned up many more such precise timers, "ticking clocks" with distinctive, regular periods. Franco Pacini and Thomas Gold (one of the proponents of the steady state universe, discussed in chapter 2) suggested that these were spinning neutron stars; however, to rotate this rapidly, these objects had to be incredibly compact. Now we know that pulsars do rotate rapidly and tick not only in radio wavelengths but also in X rays. Soon after Bell's discovery of pulsars, astronomers noticed that the star at the center of the Crab Nebula was also pulsing, about thirty times per second. Researchers had finally spotted one other type of stellar corpse.

It took more time to find the most exotic of stellar remnants: the hunt for black holes began in earnest only after astronomers and theoretical physicists who worked on general relativity finally combined forces in the late 1960s. Once again, the collaboration between theory and observations catalyzed and intensified the quest for evidence. The theorists Yakov Zeldovich and Edwin Salpeter calculated that a black hole moving around in space would feed on clouds of gas and dust that fill the regions between stars in galaxies. This, they theorized, might provide a record in the form of light, at shorter wavelengths than the visible, emitted by the heated gas and dust as they were yanked into the black hole. When a black hole swallows gas from its vicinity because of its strong gravitational pull, we say that it is accreting. Soon astronomers realized that the optimum configuration to see such an ongoing accretion event, or feeding episode, would be a binary system where a massive black hole was slowly ripping apart its partner normal star. The pull of a black hole heats up the infalling gas to extremely high temperatures, on the order of one hundred million degrees Celsius (180 million degrees Fahrenheit), the same as the sun's core. It was known theoretically that gas at these temperatures would emit radiation in the X rays. Rapid and random flickering in the X rays was recognized as definitive evidence for the presence of an actively feeding compact object—a neutron star or a black hole.

X rays emitted by swirling gas accelerated to near the speed of light by the intense gravity of black holes reveal black holes' presence. We needed detectors on telescopes that had "X-ray eyes" to unveil these extremely energetic phenomena otherwise invisible to humans. This was a challenge, because X rays, although they can penetrate human skin easily, cannot penetrate very far into the earth's atmosphere. So mounting X-ray-sensitive detectors on ground-based telescopes, akin to the way Edwin Hubble, for instance, used photographic plates to capture visible light, was futile. To open the X-ray window onto the universe, detectors somehow had to be placed above the atmosphere. This of course required technological advances in building and launching rockets—developments that were spurred by international conflicts and war, notably the Second World War.

German V-2 rockets secured during World War II came in handy for just this purpose. The first glimpse of the X-ray sky came via detectors placed in the nose cone aboard a sounding rocket flown in 1962 by the American Science and Engineering Group, led by Riccardo Giacconi. A binary star system, Sco X-1, with a neutron star orbiting a typical star, was the first cosmic object to reveal itself.

Sco X-1 blazed with an intensity that was a hundred million times stronger than X rays from the sun. An entirely new window onto the universe had now been opened. In 1970, NASA launched the first X-ray satellite, Uhuru (the name means "freedom" in Swahili, chosen to honor the support from Kenya for its launch), from Mombasa. Uhuru provided a wealth of data on the high-energy universe, finding more than three hundred individual sources, including many X-ray binaries with potential black hole and neutron star partners, as well as X-ray pulsars right in our backyard and well beyond.

Early X-ray satellites scoped out the first potential black holes, and these detections led to the blossoming of the new field of X-ray astronomy. There have been many successful X-ray satellites since Uhuru, and we have detected glowing gas swirling around nearby black holes that are several million light-years away, as well as the eddies of gas around the supermassive black holes at the centers of galaxies at cosmic distances.

The unmasking of black holes by using X rays provided the final piece of evidence verifying the theory of the life cycle of stars. For his pioneering work that "led to the discovery of cosmic X-ray sources," Giacconi won the Nobel Prize for Physics in 2002.[19] A multitude of space missions with optical cameras (the Hubble Space Telescope), infrared detectors (the Infrared Astronomical Satellite, or IRAS; Spitzer Space Telescope; and William Herschel Telescope), and X-ray detectors (the Röntgensatellit, or ROSAT; Einstein Telescope; Advanced Satellite for Cosmology and Astrophysics, or ASCA; and XMM-Newton, aka X-ray Multi-Mirror Mission or High Throughput X-ray Spectroscopy Mission) have since greatly refined our understanding of how black holes grow and evolve over cosmic time as our reach has expanded farther into the universe with more sensitive instrumentation.

* * *

The discoveries of neutron stars, pulsars, and quasars finally led to the acceptance of the idea of a real black hole. Today the entire scientific community believes in the once radical idea of black holes, and many of us are engaged in studying them and the significant role that they appear to play in galaxy formation. Some of my own research focuses on understanding the formation and growth of black holes in the universe. In particular, I am interested in the origin of the first black holes and how they became the behemoths that we see hiding in the centers of nearby galaxies. What Chandra first suggested is now conventional wisdom and the fully accepted paradigm: the first black holes are likely the corpses of the first stars that formed in the universe. But these black holes, detritus from the death of early stars that were about ten to fifty times more massive than the sun, are not expected to be particularly large. Yet within two billion years of the big bang, we have detected scores of quasars—actively feeding supermassive black holes, estimated to be a billion times the mass of the sun.

Can such rapid growth, from tiny infant or "seed" black holes to supermassive monsters, occur in such a short time? Computer models suggest that black holes would need to feast continuously on gas for the first two billion years of their lives to achieve these masses, given the conditions in the early universe.

Could we somehow make very massive, initial seed black holes? Many astrophysicists have tackled this problem, trying to derive feasible ways to do so in the very early universe. One of my postdoctoral collaborators, Giuseppe Lodato, and I have also contributed to this quest. We worked out an alternate theory and were able to show that more-massive seed black holes can indeed form from the get-go. A dramatic process—the speedy aggregation of gas in the centers of early galaxies—can produce a more massive black hole than the death of a conventional star. Astronomers call this a direct-collapse black hole. It turns out that conditions in the early universe permit such objects to form. We have been following up on the implications of such black holes, born sans star, with Marta Volonteri at the Institut d'astrophysique de Paris. We predict unique observational signatures of this kind of early black hole formation, which stand to be tested by data to be obtained from ground-based telescopes and the forthcoming NASA satellite mission the James Webb Space Telescope, due to be launched in 2018. My research group at Yale has also been avidly engaged in understanding the growth histories of the most massive black holes—ultramassive black holes, with masses in excess of ten billion times that of the sun—which have been discovered recently nearby in the universe. Pondering the question of whether black holes could keep growing unhindered and indefinitely, working with our collaborator Ezequiel Treister we predicted that theoretically there is an upper limit—a cap beyond which black holes would stunt their own growth. We made the case prior to the discovery of these behemoths. Our work claims that physical processes relevant to accretion curtail the growth of black holes and that there is a maximum limit to black hole masses in the universe.

Determining the origin of black holes is key to understanding the role that they play as they grow and glow in the galaxies that host them. The same gas supply that feeds black holes provides the raw material to form stars. The cooling of gas is crucial to the formation of stars, and the X rays produced during black hole feeding might well disrupt the supply of gas to the black hole itself, stunting its growth and heating the gas, preventing star formation. Interestingly, observations suggest the suppression of star formation in galaxies in recent times. While the details of how black holes affect their environments are yet to be fully understood, they appear to be powerhouses that could decisively alter the state of galaxies by shutting down the formation of new stars. So these unseen agents, whose very existence was vehemently challenged only eighty years ago, may turn out to play a critical role in galaxy formation. Black holes, located literally at the center of it all, may define the new map of our understanding of how galaxies assemble. Galaxies grow by colliding with one another; this inevitably implies that their central black holes might also collide and eventually merge. Astronomers can discern further properties of black holes as they merge with each other. The dying gasps of merging black holes produce an entirely new, hitherto undetected radiation, called gravitational waves. These are another consequence of Einstein's theory of general relativity. Gravitational waves, or gravity waves as they are commonly referred to, are essentially tremors in space-time, jiggles that are generated when two black holes merge, for instance. Their detectability depends on how long black holes take to merge and whether they are embedded in gas while doing so.

Many astrophysicists are engaged in the calculation of the additional observational signatures that accompany black hole mergers—any telltale signs in the X-ray, radio, or visible wavelengths. Much depends on the conditions right around the region where two black holes merge. This is a very active area of research in terms of both theoretical calculations and targeted observational campaigns to look for any such signals. It is also a problem that interests me greatly, as it combines theory and observations so beautifully. In one of the early calculations in this area, in 2002, my collaborator Philip Armitage, at the University of Colorado Boulder, and I showed that merging black hole pairs embedded in gas coalesce promptly and can therefore be detected directly by the gravitational waves they produce, as well as indirectly through radiation in other wavelengths. The discovery of gravitational waves will open a new frontier in black hole physics. The LIGO (Laser Interferometer Gravitational-Wave Observatory) experiment, currently being upgraded to Advanced LIGO, has begun taking data and is poised to detect gravitational waves from merging black holes any moment now. The gravitational wave window, expected to open soon, will offer yet another way to probe black holes, along with their optical, X-ray, and radio signatures.

The story of how black holes moved from dismissed to marginal to the center of our map is emblematic of how instruments have helped make theory real. But black holes are only a tiny part of the invisible realm. There are other enigmatic and invisible entities that rule the universe and its fate—dark matter and dark energy—which remain elusive.

4

THE INVISIBLE GRID

Coping with Dark Matter

Imagine Sherlock Holmes or Hercule Poirot tapping into their considerable inductive and deductive powers to solve a murder. There is evidence, motive, and scene, but the victim is missing. Astronomers attempting to find dark matter—an invisible substance ubiquitous throughout the cosmos—face a similar mystery. Much like when we hunt for black holes, we can only look for the effects that dark matter has on its neighborhood. We can detect its gravitational pull on objects in its vicinity and, given general relativity, the way this unseen matter bends light. Still, we face the astronomical equivalent of a prosecutor's corpus delicti—and it's harder to prove murder without the body.

But dark matter is a trickier quarry than black holes—unlike ordinary matter, it does not emit, absorb, or reflect radiation. It is inert. The only thing we know for sure is that dark matter particles, which likely formed very early in the universe, although exotic, have mass, accounting for almost all of the total matter in the universe—and that, driven by gravity, they accumulate into piles in space. All the known elements in the periodic table, including those of which we are composed, constitute a paltry 4 percent of the universe's total contents, including matter and energy, utterly insignificant compared to the amount of dark matter. Dark matter provides the scaffolding within which stars and galaxies form, aggregate, and evolve. Yet we know very little about it.[1]

Dark matter's backstory begins in an unlikely locale—a dingy, nineteenth-century glassmaker's workshop in Munich. Gaffers there blew glowing molten glass into bubbles and afterward shaped it with a blowpipe and torch. Josef von Fraunhofer, born on March 6, 1787 in Bavaria, was the eleventh and last child of the master glazier Franz Xaver Fraunhofer and Maria Anna Frohlich. There was a tradition of glassmaking going back several generations on both sides of the family. Orphaned at the age of eleven, Fraunhofer apprenticed with the court mirror and glass cutter in Munich. He was inside a workshop building when it collapsed in 1801, but he was rescued. Moved by this tragedy, Prince Elector Maximilian IV Joseph, who later became the king of Bavaria, personally provided a purse for the young boy's future. Fraunhofer used these funds to invent the tool that radically transformed all of astronomy and first revealed dark matter 132 years later.[2] Fritz Zwicky, dark matter's discoverer in 1933, would need to thank the collapse of a nineteenth-century glass shop.

The key to Fraunhofer's invention is that the light emitted by any object is akin to a fingerprint—it leaves behind specific evidence, encoded in a frequency, that identifies the unique properties of the object's chemical constituents. Bored with decorative glassmaking for the royal household, the restless young Fraunhofer took a position in Joseph Utzschneider's Optical Institute in Munich. There he received formal instruction in physics, mathematics, and optics. A quick study, he went on to write an influential essay in 1807 that demonstrates the superiority of parabolic mirrors, due to their image quality, in reflecting telescopes. Within six years of his rescue from the rubble, Fraunhofer also produced a major breakthrough in optical astronomical telescope lenses.

When light rays hit the glass surface of a lens, they bend or refract. The extent of the bending depends on the material properties of the lens (in this case, the composition of the glass) and the wavelength of the light traveling through it. In the visible spectrum, for example, red light barely alters its path through the lens, but the shorter-wavelength violet light veers. Just as the optician specifies the power of our eyeglasses with respect to a standard, we astronomers need to calibrate telescope lenses to determine the brightness of objects we observe through them. Because of the magnification of distant, faint objects afforded by the collecting area of telescope lenses, the calibration process involves developing lenses that can focus all colors together. Armed with an understanding of the wave nature of light, Fraunhofer developed the spectroscope, which splits the frequencies of light. This led to the unscrambling of the unique fingerprint of chemical elements present in a remote object's spectrum, thereby identifying its constituents. Fraunhofer's contemporaries soon recognized his talent, and he went on to become the director of the Optical Institute.

Although many of his techniques for grinding and polishing mirrors died with him, his calibration method for lenses and the invention of the spectroscope catalyzed the reshaping of our understanding of the composition and properties of all astronomical objects, near and far. His invention transformed astronomy, leading to the development of spectroscopy, the analysis of the spectrum of cosmic light sources, as a powerful and new quantitative tool. In 1812, using known light sources in his laboratory, such as sodium lamps, he determined the refractive indices for lenses and calibrated them using sunlight. Measuring the spectrum of sunlight, he detected six hundred dark lines, known today as Fraunhofer lines. Realizing that these were properties of solar light, he determined the refractive indices for each of the colors in the solar spectrum, calibrating the lens by using the dark lines as rulers. These lines reveal the atomic composition of the sun. Although Fraunhofer did not delve further into the origin of these dark lines, he measured their wavelengths and thus assembled the first spectrograph.[3] He also observed that the spectrum of the brightest stars differed from that of the sun.

Fraunhofer's first detection of dark lines in the spectrum of the sun opened the door to myriad astronomical applications of spectroscopy. Without his spectroscope, we would have had only static images—ones with no information about motion—and the incredible work of Vesto Slipher, Henrietta Leavitt, Edwin Hubble, and others would have been impossible. Astronomy would have been stuck with just pretty pictures. Quite simply, Fraunhofer set in motion the techniques and technologies that refined spectrographs, the key instrument for precision measurements of speeding nebulae, which led to the inference of dark matter 120 years later.

The arc of dark matter's story is quite different from those I have described so far. There is no mathematical theory that proposed dark matter, unlike the case for black holes, just a set of baffling observations that appeared to be inconsistent with Newton's theory of gravity in a realm where it was expected to be valid. Although astronomers inferred the masses of galaxies from their motions, upon doing so, they discovered that something was amiss. In this case it was empirical data, real measurements that simply did not make sense. The observations suggested that there was ten times more mass than what was visible. Despite strong observational evidence, scientific acceptance of the idea of dark matter, though completely data-driven was not immediate or universal. It's not surprising that astronomers resisted the idea of an invisible cosmic entity, given the fate of previously postulated invisible forces and all-permeating fluids, such as ether, miasma, and phlogiston—all eventually debunked. Invoking yet another invisible agent to explain observations was far from persuasive.

Fritz Zwicky—brilliant, creative, and cantankerous—was the

first to invoke dark matter in a paper, in 1933. He recorded the motions of galaxies in the nearby Coma Cluster in the hope that he could determine their masses. We know today that clusters are the most massive structures in the universe. All clusters, including the Coma Cluster, consist of about a thousand galaxies whizzing around at high speeds and held together by gravity. Zwicky studied the motions of the brightest eight galaxies in the Coma Cluster in detail, using the spectrograph on the hundred-inch telescope at the same Mount Wilson Observatory where Hubble had discovered the expanding universe. Zwicky found that the galaxies were all moving around much faster within the cluster than predicted if one took into account only the gravity of the visible stars. His data showed that these galaxies had speeds of about three million kilometers (about 1,864,000 miles) per hour, implying that the mass in the cluster was four hundred times denser than expected or seen. He published these results in a 1933 paper that claims that the Coma Cluster, and the entire universe by extension, must have an invisible, unseen component, "dunkle Materie," or dark matter, whose gravity likely accounts for these vast speeds.

In that paper, Zwicky opines, "If this [overdensity] is confirmed we would arrive at the astonishing conclusion that dark matter is present [in Coma] with a much greater density than luminous matter." He concludes that "the large velocity dispersion in Coma (and in other clusters of galaxies) represents an unsolved problem."[4]

Zwicky's conclusion rests squarely on the estimation of a key quantity, the mass-to-light factor, which depends on the Hubble constant. Remember that the Hubble constant relates speeds to distances in the Hubble law (as we saw in chapter 2) and can give an estimate for the current age of the universe. The mass-to-light factor is the number that defines the total light produced by a population of stars relative to their mass. In 1933, Zwicky did not dare consider challenging the value of the Hubble constant or reducing the mass-to-light ratio to resolve the discrepancy between the mass needed to explain the motions in the Coma Cluster and what he saw. Invoking dark matter was the only solution.

In 1936, three years after Zwicky published his paper, another astronomer, Sinclair Smith, made a similar case for an unseen mass component in another nearby cluster, the Virgo Cluster. Smith suggested that this missing mass might be lurking in the "internebular" voids of the cluster—in the spaces between the galaxies. Yet even after this second paper made the case for dark matter in clusters, the concept received little attention.[5]

Given Zwicky's eminence and various other seminal discoveries associated with him, it's puzzling that this initial finding did not receive the astronomy community's attention. Owing partially perhaps to his personality, progress on the issue of dark matter stalled for several decades, and the idea was resurrected only in the 1970s, to help explain a problem on an entirely different scale—the speeds of stars *within* a galaxy.[6] Forty years after Zwicky's paper reporting his discovery appeared, the astronomers Vera Rubin and Kent Ford serendipitously rediscovered dark matter while determining the masses of individual spiral galaxies. They did so by using the measured motions of stars derived by splitting their light with an image tube spectrograph and observing the red- and blueshifts of their motion. Rubin and Ford did not set out to look for dark matter—they were searching for evidence of rotation in spiral galaxies. But they found trends in the data that simply did not make sense. The motions of stars in spiral galaxies indicated that the stars were experiencing a stronger pull of gravity—and as a result moving much faster—than could be inferred from just the visible mass of stars and gas in the galaxies. For Rubin and Ford, the body was missing from the crime scene yet again. Because they were working on galaxies, though, they did not make the connection between the missing mass they required to explain the speeds of stars and Zwicky's invocation of *dun-*

kle Materie in clusters. As we have seen, the acceptance of new ideas does not come easily; the historian of science Derek J. Solla Price notes, "Perhaps it is even desirable that many of the important discoveries should be made two or three times over in an independent and slightly different fashion." This is precisely what happened in the case of dark matter.[7] The realization that it was the same elusive dark matter that could solve both puzzles on these two vastly different physical scales, a unified view, had to come from theory.

In the 1970s, theoretical understanding lagged behind observational discoveries in cosmology, but Rubin and Ford's discovery proved to be dark matter's tipping point. It took just ten years to formulate the currently accepted, dark matter–driven theoretical model for the formation of all structures in the universe. Evidence from many independent observations has established this standard explanatory model for galaxy formation, known as the cold dark matter paradigm.[8] In this model, dark matter is the primary driver of all structure formation. The existence of dark matter and the significant role it plays in the cosmos are widely accepted today. There are, however, still a few small gaps between theoretical predictions and observations. Given the sophistication of this model and how far it has been developed and honed, mounting a challenge or providing an alternative to it has been extremely difficult. But the formulation of a competing theory—an alternate view that disposes with the need for dark matter—has nevertheless been valiantly attempted, and we will soon encounter it.

One way in which dark matter's discovery and rediscovery is different from the story of black holes or of the expanding universe is that we will encounter fewer powerful individual scientists whose strong views hampered its acceptance. Science and the production of knowledge changed dramatically between 1933 and 1978—it was now a more collective enterprise, and scientific discussion had broadened. Science and technology were starting to play a more significant role in daily life. Global events including World War II and the launch of the first Russian space satellite, Sputnik, drove American science and engineering. The military-industrial complex that was set up for the war effort portended very large government investments in science and technology, which conclusively shifted the bulk of cutting-edge scientific research to the United States. Of course, crucial to all of this was the relocation of talent from Europe in the first half of the twentieth century, the rescuing of great minds that can also be viewed in part as intellectual war reparations, which began early and kept picking up pace. This also imparted a new momentum to scientific research. Critical to all of this was the intellectual leadership of the entrepreneurial astronomer George Ellery Hale. Beginning in the early 1900s, he secured patronage from American philanthropists to establish top-of-the-line instrumentation, telescopes, and research facilities on the West Coast.[9] Astronomy was poised to exploit this grand convergence.

The story of dark matter is the story of some of the most innovative astronomical minds of the twentieth century. Einstein and Hubble are also characters in this tale, while Zwicky, Rubin, and Ford are the main protagonists. Again, rivalries and competition in cosmology had sharply changed in tone by the 1970s. In contrast to prior decades, when gentlemen scientists had argued and clashed in the plush rooms of the Royal Society or Trinity College, Cambridge, now there was broader participation in science, in particular as the center of gravity of new discoveries shifted slowly from continental Europe and the United Kingdom to the United States. There were open debates at specially convened meetings, and the field became more inclusive and more global. It was the start of the democratization of astronomy. On the intellectual side, one could say that astronomy was maturing as a science, as it evolved from focusing on the detection of individual objects and phenomena to increasingly preoccupying itself with more systematic and precise measurements. At this time, there was a growing demand and need for higher-precision instruments to make and replicate observations with increased accuracy. The case of dark matter illustrates the new and critical role that instruments and observations played in stimulating the development of a theory to explain data. Reluctance to accept the idea of dark matter illuminates many other new dimensions of the practice of science, which had evolved significantly since the early 1900s.

In the children's fable *The Little Prince,* the fox tells the prince, "The essential is invisible to the eyes." But the human preoccupation and fascination with the unseen date back to well before Antoine de Saint-Exupéry's 1942 novel. Early modern science often assigned invisible causes to unexplained phenomena. Natural philosophers supposed invisible agents to be the causes of disease, the medium for light's propagation, and the fuel for the burning of substances. Before the development of the concept of germs, many believed that the inhalation of a miasma—literally an unhealthy smell—caused disease. One seventeenth-century theory held that substances are able to burn when they contain a firelike element, phlogiston (objects that could burn in air were believed to be rich in phlogiston), and that the burning stopped when the air was unable to absorb any more of this substance. Robert Boyle, whom many consider the first modern chemist, was one of the first to suspect that air was not one substance but a mixture of many. In his 1674 *Suspicions about Some Hidden Qualities in the Air,* he says of air, "There is scarce a more heterogeneous body in the world." His examination and study of the role of air in the processes of burning and oxidation finally led to the rejection of phlogiston.[10]

The idea of unseen effluvia attracted proponents even in the nineteenth century. Until Albert Michelson and Edward Morley's 1887 experiment disproving it, many believed that an all-pervasive medium, ether, existed and enabled the propagation of light waves and gravity. In the context of Newton's corpuscular theory of light, ether was thought to be the medium that facilitated the transport of light particles from the emitter. This belief stemmed from an analogy with sound: it was known that sound waves propagate by squishing their medium—air—compressing and releasing air particles, transmitting vibrations that transport finally to our eardrums. The presence of a medium was therefore seen as essential to the propagation of light waves too—hence the postulation of ether. If ether did fill space, then the motion of the earth through it while orbiting the sun should be detectable and measurable. Michelson and Morley set up an experiment to measure this motion. They used what is called an interferometer, which in a way races two beams of light against each other. One had to travel against the proposed current of ether, the other perpendicular to it. If the current was real, there should have been a noticeable disparity in the time that light took to travel along these two paths, but Michelson and Morley found none. Ether did not exist.[11] This is one of the most famous "null result" experiments in physics to earn a Nobel Prize. We know now that light travels as an electromagnetic field—it requires no medium to propagate and is fastest, in fact, in a complete vacuum. This same principle of interfering waves is what the LIGO experiment (described in the previous chapter) relies on, although the waves in question there are gravity waves, tremors generated in space-time when two black holes merge. In that case the two path lengths would vary, as gravity waves would change the length of the two arms in the experiment.

With Einstein's formulation of general relativity it became clear that the force of gravity also requires no mediation—it manifests locally as divots around objects with mass in the fabric of four-dimensional space-time. Amid the new and growing understanding of the cosmos in the 1920s and 1930s, the major breakthrough, as we saw earlier, was Hubble's discovery of the expanding universe. This discovery was a consequence of measuring the distances of extragalactic nebulae by using Cepheid variables, stars whose properties enable the inference of accurate distances. As Hubble and others measured the distances and speeds of galaxies external to our own, still others hoped to use Newton's laws of gravitation—believed to hold true throughout the universe—to go further and determine the masses of these galaxies.

Intellectual reputations in science are usually built gradually, but Hubble reached the pinnacle rapidly. By the early 1940s he was at the zenith of observational astronomy and his work beyond question. There was a silent rivalry between him and Zwicky, as they were both at Caltech (the California Institute of Technology) and vying for the same telescope resources. Hubble always had the lion's share of resources and telescope time, which, understandably, made Zwicky unhappy. Of course, the instruments and technology that would enable more precise measurements and challenge the value of the Hubble constant were yet to be built. So because Zwicky's proposal of *dunkle Materie* was radical and speculative, it did not provoke a reexamination of Hubble's work. Indeed, Zwicky himself found his case for dark matter unconvincing and remained a bit skeptical, like Hubble with the expanding universe. Even as late as 1957, Zwicky admitted, "It's not clear how these startling results [from the Coma Cluster] must ultimately be interpreted." The idea of yet another dark, unseen, and invisible entity was hard to take seriously, even for the person who first suggested it.[12] As we have seen, the original proposers of radical scientific ideas are themselves often resistant to accepting them or their implications. The far-reaching consequences of these ideas are usually the cause of this struggle.

Although his 1933 paper had gone unnoticed by the astronomy community, Zwicky was not giving up—he explored its idea further. He realized that if clusters contained a large amount of invisible matter, it should curve space-time. As light rays traversed the divot

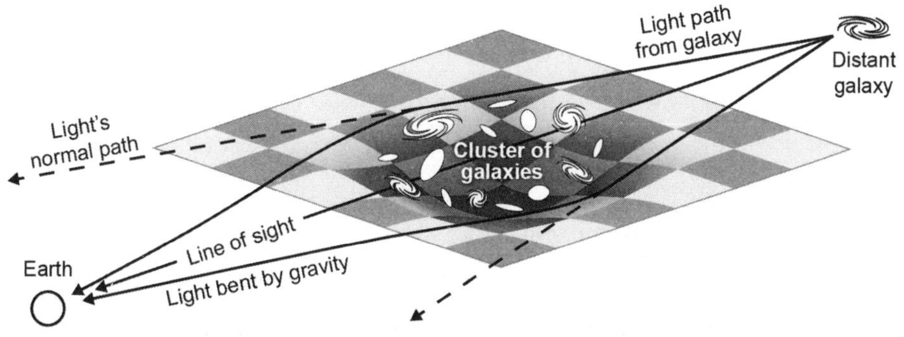

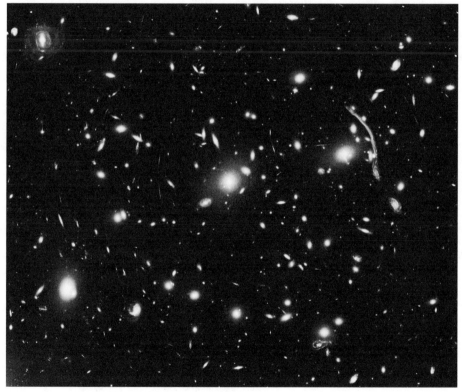

Top: Bending of light rays and the pothole in space-time caused by a cluster of galaxies.
Bottom: Hubble Space Telescope image of gravitational lensing detected in the galaxy cluster Abell 370, originally observed from the ground-based Canada-France-Hawaii Telescope. Courtesy of NASA, ESA, the Hubble SM4 ERO Team, and ST-ECF.

in space-time induced by the enormous gravity of the cluster, they would be deflected from a straight path. In other words, a cluster would act like an optical lens, diverging and converging light rays. Zwicky referred to these massive clusters as gravitational lenses. In a paper published in 1937, he made the case for dark matter again, estimating the light deflection by a cluster and noting that it was an inevitable consequence of the large amounts of dark matter in clusters but would be too small for instruments at that time to detect.[13]

Given the lack of appropriate instruments, Zwicky's proposal did not garner much attention until the late 1960s. At that point there was a resurgence of interest in and extension of his work on the predicted amount of light bending, by the astronomers Sjur Refsdal, Ramesh Narayan, and Roger Blandford. They realized that there are particular circumstances in which lensing effects can be extreme and therefore more easily detectable. They found that when background galaxies are perfectly aligned behind massive clusters, the light they emit is on occasion so severely stretched out into long arcs that these sometimes split into two. What this cleavage of the light ray does is produce two magnified images of the same background galaxy. Depending on the precision of the alignment, an even larger number of virtual image copies of the background galaxy might appear. For instance, in the Hubble Space Telescope image of a very massive cluster, CL0024+16, which acts as a lens, the same background galaxy appears five times over! We know that these are images of the same object and not just astronomical doppelgangers because we can measure their spectra, their unique chemical fingerprint. The spectra of these five images are all identical.

Another notable feature of these multiple images is that some of them can be highly stretched, so a regular background galaxy with an oval shape might get mangled into many copies, some of which will be misshapen into rather elongated ellipses or arcs. Dramatic arcs produced by strong lensing, as theoretically predicted, are now seen routinely in high-resolution images of clusters. In cases where the alignment of the background galaxy and the cluster is less perfect, light rays from the galaxy are very slightly bent, causing a gentle stretch referred to as weak lensing. When the astronomers Geneviève Soucail and Bernard Fort detected a highly elongated arc in the cluster Abell 370 with the Canada-France-Hawaii Telescope in Hawaii in 1987, they knew that unless they had a measurement of its spectrum that matched the spectra of other copies of the same galaxy, they could not convince anyone that this was gravitational lensing in action. Taking the spectrum and finding that it was identical to those of the less distorted copies in the image, they were able to prove that they had detected gravitational lensing.[14] The improved optics of telescopes had finally proved Zwicky right.

Since then, the exquisite resolution of the Hubble Space Telescope has allowed the detection of many more gravitationally lensed galaxies. Tracing the path of light rays, we can now reconstruct the detailed distribution of the unseen matter in clusters that causes this deflection of light. Lensing clusters, however, are rare objects; much of the night sky is undistorted and discloses the true shape of galaxies. These undistorted shapes provide a baseline to determine the strength of lensing produced by regions peppered with galaxy clusters. My own expertise lies in mapping the concentrations of dark matter in clusters that produce these lensing effects. Using Hubble Space Telescope data of several lensing clusters, my work and that of many others in the field has revealed that dark matter dominates clusters on both large and small scales. In fact, it is aggregated dark matter that is responsible for these lensing effects, because the mass of visible stars in the clusters' galaxies is insufficient to produce the strength of the observed deflections. Mapping the dark matter inside clusters, we find that it consists of two kinds: a smoothly smeared one and a clumpy one, which can be associated with galaxies in the cluster that are bound together by gravity.

Lensing is a ubiquitous phenomenon, in the sense that all masses cause deflections in light ray paths because of the divots and valleys that mass imprints on the fabric of space-time. The more massive the lens, the more dramatic its lensing effect and the more readily observable it is. Gravitational lensing offers a unique and independent tool for measuring important properties of our universe, and as we will see in the next chapter, it can even come in handy for exploring the nature of yet another invisible and mysterious component of our universe, dark energy.

There were a couple more thwarted resurrections of the idea of dark matter before it began to be taken seriously. In fact, it was repeatedly invoked and then ignored or discarded until yet another set of observations needed its explanatory power. Puzzlingly, most of these attempts were independent and done with no knowledge of prior work. This continual reemergence followed by rejection and forgetting of the concept of dark matter is typical of the life cycle of extremely radical scientific ideas.

The next milestone in our story came from observations on a smaller scale: the scale of individual galaxies—as opposed to clusters—which by the 1930s were more familiar objects and were observed to be more numerous in the universe. In 1940, the well-respected Dutch astronomer Jan Oort claimed, after detailed study of a spiral galaxy, that "the distribution of mass in this system appears to bear almost no relation to that of its light."[15] This was again a bold assertion, as there was no reason to believe in the existence of anything other than visible stars as the constituents of galaxies. Hubble's mass-to-light factor was still beyond reproach.

The next resurgence of the idea came in 1959, with observations closer to home, when the astronomers Franz Daniel Kahn and Lodewijk Woltjer inferred the mass of our galaxy and its nearest neighbor, the Andromeda Galaxy. Unlike every other galaxy in the universe, they found that Andromeda is hurtling toward rather than away from the Milky Way, and this, they concluded, meant that there is gravity from some unseen mass at play. Since a full accounting could be made of all the visible stars, they claimed that most of the mass had to be in some invisible form. Yet again, they made no connection to Zwicky's prior assertion for just such a component in the Coma Cluster of galaxies. In fact, Kahn and Woltjer seem to have been completely unaware of Zwicky's and Smith's previous work on unseen mass in nearby galaxy clusters, because they do not cite these papers in their work.[16] Meanwhile, Zwicky was soldiering on, slowly undertaking the study of more nearby galaxy clusters to test if these structures were indeed held in balance, as the need for dark matter hinged on this assumption. Despite doggedly pursuing this angle, he seems to have abandoned his idea of missing mass and the search for *dunkle Materie*. It took longer, however, for astronomers to understand that the same dark, invisible component could explain the motions measured in objects on different physical scales: clusters and galaxies. In the thirty-some years after Zwicky's original speculation in 1933, they had begun to discover a plethora of dark, compact astrophysical objects—namely, neutron stars and black holes, which do not emit light, unlike stars. Could galaxies and clusters both be filled with an abundance of these inherently dark objects? Could these be the dark matter whose gravity was being detected? In time, researchers considered but ultimately rejected the possibility that these dark objects could be dark matter.

After Kahn and Woltjer's brief exploration, the entire subject of dark matter was ignored for a good while. To examine why this idea lay dormant and discarded for so long, it's worth looking more closely at the key player—Zwicky. He finished his PhD at the Swiss Federal Institute of Technology in Zurich in 1922, working on ionic crystals and not astronomy. He continued his research at his alma mater for three more years before departing for the United States. At this juncture, U.S. science was very international, and the country was an attractive destination for ambitious European scientists. There were new philanthropic programs that enabled the recruitment of talented young scientists from abroad. The postdoctoral fellowship program of the International Education Board of the Rockefeller Foundation was one of the first, running between 1924 and 1930 and attracting 135 European physicists. This was followed by the emigration of about eighteen hundred scientists, mostly Jewish, seeking refuge from the Nazis. Zwicky was part of the first wave of émigrés and was the recipient of a Rockefeller postdoctoral fellowship. In 1925, at the age of twenty-seven, he went to Caltech to work with the experimental physicist Robert Millikan. Zwicky arrived at an opportune time. Hubble's pioneering observational work was under way, and planning for the next generation, two-hundred-inch telescope on Mount Palomar had begun.

Soon, exclusive access to the largest telescope in the world (the Palomar telescope superseded the one on Mount Wilson) spurred Zwicky and several other scientists to shift their focus to astronomy and astrophysics. Zwicky's switching fields proved to be a savvy and intellectually fruitful move. As a foreign-trained outsider, he offered a fresh perspective and initiated many creative projects. But he was an extremely short-tempered, abusive, and opinionated person, with an abrupt and dismissive style that rankled his colleagues. He had been trained in a vastly different, more hierarchical academic culture than the one he now found himself in. Many had trouble tolerating his sense of self-importance, though he did have admirers. A 1974 tribute from a colleague praised him nonetheless: "Zwicky possessed that necessary concomitant of greatness, the generation in others of a strong positive or negative response. . . . Those who see further or deeper are not universally admired."[17]

Zwicky's brusque manner, coupled with his arrogance, likely adversely impacted the attention paid to his work. He was brimming with ideas; many were wrong, but several turned out to be right, including those that were ignored. In *Lonely Hearts of the Cosmos*, Dennis Overbye describes the scientific community's lack of faith in Zwicky: "[He] had so many ideas it was almost impossible for other astronomers to sort the good from the off-the-wall."[18]

As we have seen, the personalities of scientists often strongly impact how the rest of the community receives their work, regardless of its significance. The culture of hero worship and the collective recognition of genius often cause colleagues to exempt brilliant scientists from socially accepted codes of behavior. So one who is deemed a genius is often given a free pass, and his bullying, brazen behavior is overlooked, but not always. Zwicky was one of those scientists not granted that immunity. Therefore, given his irascible temperament, his proposal of dark matter languished for an unusually long time.

Yet scientists could not ignore it forever. In 1970, Rubin and Ford started working together on galaxy dynamics at the Carnegie Institution's modest astronomy program in Washington DC. Rubin, a slight, soft-spoken, and determined woman, is one of the most eminent astronomers alive. She was not a person who courted controversy, so she and Ford held off on sharing their results that suggested the need for vast amounts of invisible mass in the spiral galaxies they surveyed. They cautiously published papers reporting their curious data and offering many other explanations, clearly skirting the dark matter interpretation. They concluded a 1973 paper written with Judith Rubin (Vera's daughter) with a sentence that clearly deflects attention away from what was at stake in their research: "Obviously we are not through with this business."[19] Rubin and Ford did not make the connection between their findings in distant spiral galaxies and those from spiral galaxies in our own backyard—the Milky Way and Andromeda—reported by Kahn and Woltjer in 1959. In fact, they too seem to have been unaware of all earlier work on dark matter: Zwicky's inference from clusters of galaxies remained unknown to them as well.

As a woman, Rubin had an atypical path to astronomical research. She enrolled at Cornell University in 1950 for a master's degree to join her husband, who was pursuing his doctorate there. Rubin's master's thesis project involved searching for any systematic motion within galaxies; in particular, she was looking for rotations. Pure curiosity motivated her work, because there was no theoretical basis for understanding whether galaxies rotate or not. In a way, her position at the margins of the profession gave her the freedom to ask fresh questions that perhaps might not have been encouraged at Princeton, Harvard, or Caltech, which were the traditional citadels of astronomy at the time. She presented her findings at the 1950 American Astronomical Society Meeting, in Haverford, Pennsylvania. In an interview for the American Institute of Physics in 1996, Rubin recounted that she had given birth to her first child a few weeks previously and that she nervously walked into the room, not knowing any of the assembled greats there. Her talk was rather grandly titled "Rotation of the Universe," but that was a naïve, not arrogant, choice. The response was extremely hostile, and the general tenor of comments was that one simply could not do what she was attempting. But amid this skepticism, she vividly remembers one mild-mannered man with a strong German accent who gently encouraged her, saying "that this is an interesting thing to do, that the data probably are not good enough, but that it was an interesting idea for a first step." This gentle encouragement, which made her "feel somewhat less than mashed to the ground," came from none other than Martin Schwarzschild, an expert in the dynamics of galaxies and one of the pioneers of computational astrophysics, who had worked on the Manhattan Project.[20] He was also the son of Karl Schwarzschild, whom we encountered briefly in chapter 3 as the discoverer of the mathematical solution to Einstein's equation that corresponds to a black hole.

Even though Rubin retitled her paper to the more modest "Rotation of the Meta-galaxy," both the *Astrophysical Journal* and the *Astronomical Journal* rejected it. She recalls that part of the objection to her work was astronomers' belief that the concept of large-scale motions within galaxies was rather ridiculous. It was hard to reconcile such internal motions with the overall expansion of the universe. Undeterred, she moved on to graduate work at Georgetown University under the supervision of George Gamow, one of the founding fathers of the big bang model. Rubin abandoned her work on these large-scale motions and rotation, partly because it would have been primarily observational. She had two young children by then and did not think it feasible to take on an intense project that would require frequent travel to far-flung telescopes. Besides, after the controversy over her master's thesis work, she knew that she did not enjoy being in the midst of a storm. She therefore decided to embark in a completely different direction, trying to understand if there were regularities in the distribution of galaxies in the sky.

By the time Rubin had completed her PhD and was working at the Carnegie Institution, she was again looking at the motions of stars in galaxies. She had teamed up with her colleague Kent Ford. He had built an incomparable instrument, the most sensitive spectrograph available at the time. They used it to study starlight originating from many different parts of spiral galaxies. They looked at the light produced in the dense centers and in the sparser outskirts of these galaxies. The stars that make up the disk of a spiral galaxy revolve in circular orbits around the center. If the disk is inclined even slightly with respect to us, its stars appear to be moving toward us along one side and away from us on the other. As described earlier, when a source of light moves toward us, we see a decrease in its wavelength, which produces a shift to the bluer end of the visible spectrum. Similarly, light from stars that appear to be moving away from us changes wavelength to the redder end of the spectrum.

This shift in wavelength of light (or sound) waves, the Doppler effect, is due to the relative motion of the source with respect to the observer. Rubin and Ford measured the Doppler shifts across the disks of several spiral galaxies and, using this data, calculated the orbital speeds of stars at different locations within those galaxies. They plotted a graph of the speeds of stars relative to their distance from the center of the galaxy. It is reminiscent of Hubble's graph of the motions of galaxies with respect to their distance from earth, except Rubin and Ford were focused on the motions of stars held inside individual galaxies by their gravitational grip.

What they saw was most peculiar and unexpected. To understand what was so odd, let's first focus closer to home, on the motions of planets around the sun. In our solar system, where the sun's gravity dominates, the inner planets move faster in their orbits than the outer ones. The farther from the sun—the most concentrated mass in the solar system—the more slowly a planet moves, taking significantly longer to complete one full revolution. This is because the force of gravity exerted by the sun weakens with distance, so the outer planets feel a much diminished inward pull. For example, the force of gravity at a location twice as far from the sun is four times weaker. It is the slower motion and not just the size of the orbit that increases the time it takes to complete one circuit. Mercury, for instance, takes eighty-eight Earth days to go around the sun, Saturn takes twenty-nine years to make one full revolution, and Pluto requires nearly 250 years. Looking at a similar relationship in spiral galaxies, Rubin and Ford found just the opposite—the speeds of stars appeared to be much faster the farther away from the center they lived. They also appeared to reach a peak value and then remained unchanged. This was an extremely strange finding, contrary to what one would expect on applying Newton's laws, assuming that the gravity was provided only by all the visible stars. Rubin and Ford found this trend in all the spiral galaxies they observed. There was only one plausible ex-

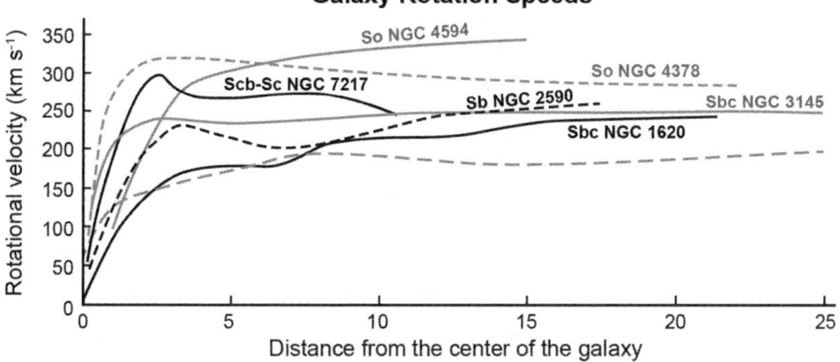

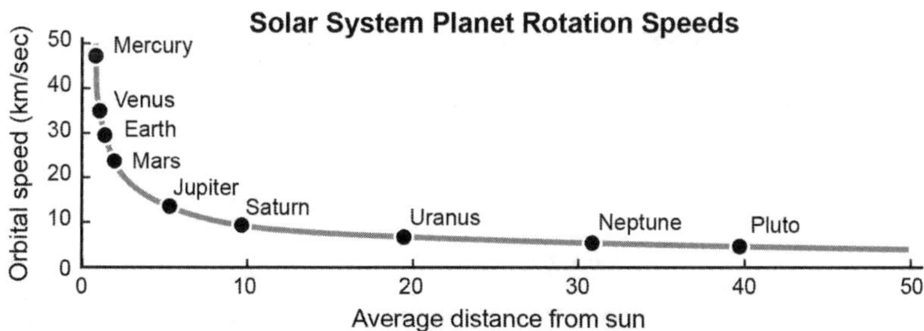

Top: Rotation curves of stars in galaxies based on Rubin, Ford, and Thonnard's paper "Extended Rotation Curves of High-Luminosity Spiral Galaxies. IV. Systematic Dynamical Properties Sa→Sc," Letters to the Editor, *Astrophysical Journal* 225 (November 1978): L109, fig. 3.
Bottom: Rotation speeds of planets in the solar system in kilometers per second. A comparison of these two graphs provides evidence that we need additional gravitating matter (dark matter) to explain the observed motions in galaxies.

planation: there was a considerable amount of unseen mass in the outskirts of galaxies—mass that did not emit light and was hence not part of the inventory of gravity inferred from the observed starlight. In essence, some mysterious ingredient seemed to keep the stars afloat at the same speed in the outer reaches of galaxies. Once again, for this interpretation the value of the mass-to-light factor was key. As we noted before, this number depends on the value of the Hubble constant, which, thanks to more data and more accurate measurements, had been revised since Zwicky's first postulate of *dunkle Materie*. Nevertheless, this update still did not alleviate the need for dark matter. It finally seemed to be here to stay.

In 1975, Morton Roberts and Robert Whitehurst expanded Rubin and Ford's work by measuring the speed of gas in the outer parts of galaxies, regions where stars have become much sparser. Probing the relationship between distance and speed beyond the region where there are visible stars, they found that the speeds stayed constant and the effect of potential invisible mass at the outskirts persisted.[21] These findings, which required vast amounts of unseen mass to make sense, were contested and received with skepticism at conferences and meetings.

Fierce debates ensued among astronomers, and there was much discussion of how the invisible dark matter was distributed in these galaxies. Despite consensus among Rubin and Ford and Roberts and Whitehurst about the existence of unseen matter, they had yet to make the connections with Zwicky's and Smith's earlier work on dark matter in clusters or with Kahn and Woltjer's claim for dark matter in the Milky Way and the Andromeda Galaxy. What finally catalyzed the associations between observations on different physical scales came inadvertently from theory.

In 1973, Jeremiah Ostriker and James Peebles, young theorists at Princeton who were working on a related, theoretical question about the stability of galaxies and their disks of stars, suggested that dark matter could be useful in keeping galaxies anchored. It is interesting to note that this paper does not refer to any of the observational work and really came as an independent theoretical calculation concluding that "masses of our Galaxy and of other spiral galaxies exterior to the observed disks may be extremely large."[22]

The following year, Ostriker, Peebles, and Amos Yahil published a paper on the mass distribution in galaxies from the center outward that convinced most of the community that the missing mass was real and, what's more, that it played a crucial role in holding all galaxies together. Their work shows that an extended distribution of invisible matter, now dubbed a spherical halo, holds the stars tightly in their galaxies. Ostriker, Peebles, and Yahil concluded that there is significant mass in the outskirts of our Milky Way and other spiral galaxies. Few accepted the idea of ubiquitous dark matter initially, and some detractors argued that galaxies could be held steady by other means—perhaps a "bulge," a surfeit of stars in the inner region. The gravity of this bulge, they contended, would be sufficient to hold galaxies together, and in fact, the presence of a dark matter halo would suppress the formation of spiral structures in galaxies.[23] Today, observations strongly support Ostriker, Peebles, and Yahil's conclusion that the invisible mass not only extends to the outer regions of galaxies but is significant everywhere in the cosmos.

It also took time for astronomers to comprehend that the missing mass needed to explain motions in galaxies was the same as the missing matter needed to explain the motions of galaxies in clusters and the light deflections they produced. In 1961, Viktor Ambartsumian, an astronomer at the Yerevan Observatory in Armenia, first suggested a connection between these physical scales—that the invisible matter inferred in clusters by Zwicky and that in spiral galaxies was likely the same. The acceptance of this idea, which conjoined puzzling observations on two scales, was slow to come. In fact, the first conference dedicated entirely to dark matter was held in Tallinn, Estonia, only in January 1975. However, at this conference there were many lively debates not on whether there was a strong observational case for dark matter but rather on potential dark matter candidates, which included several plausible possibilities: ionized gas, faint stars, and collapsed objects such as neutron stars and black holes. Finally, the discussion revolved around candidates that might constitute dark matter in both galaxies and clusters. In addition to the obvious ones—which did not produce light, like black holes—

conference participants explored a more exotic suggestion: perhaps unseen matter was composed of particles fundamentally different than those of ordinary matter. Once it was accepted, by the late 1970s, that the dark matter problems in galaxies and in clusters were one and the same, and real, it became clear that dark matter might play a significant role in every kind of galaxy and throughout the universe. Theoretical calculations of how mass in the universe aggregates and assembles to form the structures, such as galaxies, that we do see suggest that dark matter particles are cold—that is, their motions are slow and fairly unenergetic. Thus, we came to recognize this unseen, sluggish, but ubiquitous component as cold dark matter.[24]

A paper by George Blumenthal, Sandra Faber, Joel Primack, and Martin Rees published in *Nature* in 1984 lays out the framework for the formation of galaxies and clusters in a cold dark matter–dominated universe. Around the same time, X-ray studies made clear that elliptical galaxies also have missing mass. This growing empirical evidence aligned well with early numerical simulations in which dark matter appeared to drive the formation of all of the universe's structure—spirals, ellipticals, and clusters. But despite knowing what it could do, researchers still had no clue what dark matter was. They considered diverse candidates, ranging from compact objects—such as black holes, brown dwarfs (stars that failed to ignite, so they have mass but do not produce light), and white dwarfs—to gas. Even neutrinos, ghostlike particles that barely interact with most matter, auditioned for the role. Yet in 1983, computer simulations designed to test whether neutrinos could be a candidate for dark matter failed to reproduce the observed properties of galaxies in the universe. One by one, most dark matter candidates were tested and ultimately discarded. A few viable candidates have survived and remain contenders, but the dark matter particle was and remains elusive. In a review article for *Science* written in 1983, reminiscing on her work on the rotation curves of spiral galaxies, Rubin noted, "Astronomers can approach their tasks with some amusement, recognizing that they study only the 5 or 10 percent of the universe which is luminous."[25] Either dark matter is truly made of exotic stuff unlike ordinary matter, or astronomers must question Newton's laws of motion. The staunch belief that Newton's laws are applicable to individual galaxies allowed for the dark matter hypothesis and therefore a more comprehensive role for dark matter in the universe. But what if astronomers had challenged Newton? What if they had just said, in the case of dark matter, that classical laws of gravity need not apply at large cosmic distances? After all, there was clearly a precedent—Einstein had upended Newton when it came to the nature of gravity.

Entire theories are not usually discarded easily; new observations that do not fit an existing paradigm more often than not lead to amendments of the currently accepted world view rather than an entirely new one. This is the course of "normal science," as pointed out by the historian and philosopher of science Thomas Kuhn in *The Structure of Scientific Revolutions*.[26]

Consider for example Sir William Herschel, the English astronomer who, with a telescope of his own making, discovered the planet Uranus on March 13, 1781. He pushed the known edge of the solar system beyond the classical planets with his sighting. By 1846, Uranus had almost completed a full orbit since Herschel's first observation of it. Astronomers, tracking its orbit, found discrepancies that Newton's theory of gravitation could not explain. This pointed to the possibility that either Newton was wrong or his laws of motion needed amending. Starting with the observed anomalies, the French astronomer Urbain Le Verrier suggested that an undetected planet lurking beyond Uranus was affecting its motion, and he calculated where such a planet ought to be. His prediction was verified when Johann Gottfried Galle and Heinrich Louis d'Arrest detected Neptune on September 23, 1846. The British astronomer John Couch Adams was also on the trail and had made an independent prediction; however, Le Verrier beat him to the chase and was the first to report the detection of Neptune. Newton's laws remained intact.[27]

Yet Mercury too seemed to depart from Newton's laws. Given his previous success, Le Verrier suggested that there might be another hidden planet causing this odd orbital behavior. Long and unsuccessful searches even led to false claims of finding such an unseen planet, dubbed Vulcan. But there is no Vulcan. In the end, it was Newton's laws that needed an overhaul in this instance. In a 1916 paper titled "The Foundation of the General Theory of Relativity," Einstein used his new theory to correctly predict the precession in the orbit of Mercury.[28] As we saw in the previous two chapters, his theory of general relativity, published in 1915 and 1916, supplanted Newton's theory of gravitation.

As in the case of Newton and Einstein, sometimes when observations do not fit a current theory they are harbingers of a whole new one, but most often they simply illuminate a neglected or incomplete detail in the established model. The vast majority of astronomers today believe in the existence of dark matter, because even though we have yet to detect the particle responsible for it, the evidence from astronomical observations of the motions of galaxies and the bending of light by clusters is overwhelming. Those who believe in dark matter do so because of corroboration from many independent lines of indirect evidence. Besides, simulations of the assembly of galaxies and clusters show that webs of dark matter pervade our universe and that galaxies form where these filaments intersect.

Resistance to the idea of dark matter, on the other hand, has led a small group of physicists to question the fundamental laws of gravity. Arrigo Finzi proposed this possibility in 1963, in a paper published in the *Monthly Notices of the Royal Astronomical Society*. He revisited Zwicky's observations of the motions of galaxies in clusters and tried to explain them by assuming a new law of gravitation, one that implies a stronger attraction at larger distances instead of the weaker pull predicted by Newton's laws. In doing so, he dismissed the unseen: "If one accepts the ideas suggested in this paper, one has no particularly strong reason for suspecting the presence of a very large amount of invisible matter." He concluded by stating that rapidly accumulating data might settle the matter in the near future.[29]

Finzi's suggested modification requires us to understand how Newtonian gravity acts. According to Newton's theory, the strength of gravity's pull falls off with distance. The farther masses are from one another, the weaker their mutual, attractive pull. This form of the law holds well within our daily experience on earth and with slight corrections from Einstein's theory of general relativity appears to hold within the solar system. But what if we go farther out into the cosmos?

Inspired by Finzi's paper, the physicists Jacob Bekenstein and Mordehai Milgrom wondered if gravity could be different on cosmic scales, in the regime where accelerations induced by gravity are very small. They suggested a theory that modifies the laws of gravitation under such conditions, called modification of Newtonian dynamics (MOND).[30]

Although the evidence for the existence of vast amounts of dark matter is secure and mounting, it all relies on our interpretation of the data, since the dark matter particle is still missing. According to MOND, when the acceleration due to gravity falls below a certain value, gravity's pull no longer decreases as quickly. Instead, it grows. In the case of stars orbiting a galaxy, acceleration and distance from the galactic center are linked, so gravity is modified and is stronger than Newton's prediction far from the center of galaxies. The net effect is that stars on the outskirts of a galaxy orbit the center with the same speed as stars closer in, as observed. Observationally, MOND succeeds in explaining the motions of stars in galaxies—and works best for the population of faint galaxies—but it fails to explain clusters, where the need for dark matter first arose.

The one place where MOND fails spectacularly is in explaining the observed and confirmed light deflection effects produced by clusters. Even if Newton's laws have been modified, Einstein's theory of general relativity still needs to hold, because we observe gravitational lensing effects. Mass is needed, according to general relativity at least, to curve space and affect the paths of light rays. To explain the light bending seen in clusters, we need a large amount of invisible mass, which is responsible for producing a deep divot. MOND, it turns out, also needs to invoke an unseen extra mass component to explain lensing, and some researchers have made the case, once again, for those diminutive particles neutrinos as the dark matter in clusters. This wrinkle, the inability to match the observed lensing data without adding in neutrinos, makes MOND less convincing and less appealing. By modifying gravity, it dispenses with the need for dark matter in galaxies, but it can't explain away the need for dark matter in clusters. The cold dark matter theory still prevails. Data from the Hubble Space Telescope has shown that lensing by galaxies and clusters is not uncommon in the universe, and all current observations are entirely consistent with theoretical predictions of how dark matter is clumped and smeared in the universe.[31]

All the spectacular successes of the cold dark matter theory cannot be easily replicated by MOND. Its key shortcoming is that it is not really a comprehensive theory, such as Newton's or Einstein's; it does not yet offer any fundamental physical reason for the proposed alteration of gravity but just aims to match the observational data. The existence of a deeper, underlying theory that might give us the modification of gravity that MOND posits also remains elusive. If such a theory did exist, it would need to explain all the current observations, fill all of dark matter's roles—forming the structure of the universe, expanding the universe, and lensing light—and make new testable predictions in order to supplant the dark matter hypothesis.

For any new theory to displace an old one, it must explain all existing data and make additional testable predictions that we can verify observationally. For two competing theories to go head-to-head, they both have to explain existing data and make verifiable predictions. So although MOND is not yet a fully viable alternate theory, it might offer a glimpse of an alternative theory of gravity. MOND provides an ongoing and active field of research, although only a handful of astronomers are testing it and few theorists are working on refining the formulation. Despite this, the debates over MOND versus cold dark matter can become intense. Cold dark matter theory has tremendous explanatory power, yet there are a few gaps, instances where it fails to match observations perfectly. The few situations of tension between this theory and observations occur where baryons (ordinary atoms) jostle close to dark matter particles, such as in the innermost regions of galaxies. In the centers of galaxies, where stars are tightly packed and baryons outnumber dark matter particles, the cold dark matter model fails to describe observed properties accurately. Distinguishing the roles of dark matter and ordinary atoms in these crowded cosmic corners has been challenging for both observations and numerical simulations.

It appears that the entire universe is smeared with dark matter, a cosmic web with a distinct filamentary structure that permeates even the spaces between galaxies. We now have precision maps of dark matter derived from gravitational lensing observations. The most recent, highest fidelity ones produced by my research group, from cluster lensing data obtained from the Frontier Fields project, reveal the existence of what appear to be dark matter halos around tiny dwarf galaxies inside clusters that are five billion light-years away. Light deflection has allowed us to quantify the presence of dark matter associated with the smallest cluster galaxies in the universe. And dark matter seems to exist on a variety of scales in the universe. Still, it is worthwhile to ask if and why the nature of gravity remains unchanged on cosmic scales. Most helpful, of course, would be to find the putative dark matter particle, the missing body at the crime scene. In terms of candidates to consider, we have explored the entire gamut, from ordinary matter—planets, faint stars, and black holes—to the exotic. Cosmologists refer to ordinary matter candidates collectively as massive compact halo objects (MACHOs). Theory now tells us that if dark matter is no different from ordinary atoms and particles, then there simply isn't enough total matter to go around. We can calculate how many ordinary atoms were created at the big bang, and observations of radiation left over from it confirm that estimate. From looking at the mass budget of the universe, it is clear that we need some kind of exotic particle, created in the early universe, that is distinct from ordinary matter to account for all the inferred dark matter. Of course, such particles would, by definition, be difficult to detect—fairly lazy and barely interacting with normal/ordinary matter. These weakly interacting massive particles (dubbed WIMPs by cosmologists) would easily pass right through you. There are many ongoing experiments to directly detect dark matter particles, WIMPs that stray close to earth. So far, however, this mysterious, ubiquitous particle has eluded detection.

Unraveling the role of dark matter in the universe marks the beginning of a new chapter in cosmology. The practice of science has gradually evolved in the past sixty years—requiring collective teamwork and an arsenal of new tools. Powerful computers with superior graphic capabilities now allow us to follow the evolution of the universe and visualize it, enabling direct comparisons with astronomical observations. One of our key limitations as cosmologists arises from the fact that unlike other scientists, we cannot perform controlled experiments. What you detect is what you get! Cosmology, originally a speculative discipline, is now respectable, as numerical simulations have become detailed surrogates for experiments. By the 1980s, cosmology had three modes of inquiry, three independent approaches that are crucial to the production of new knowledge and the test-

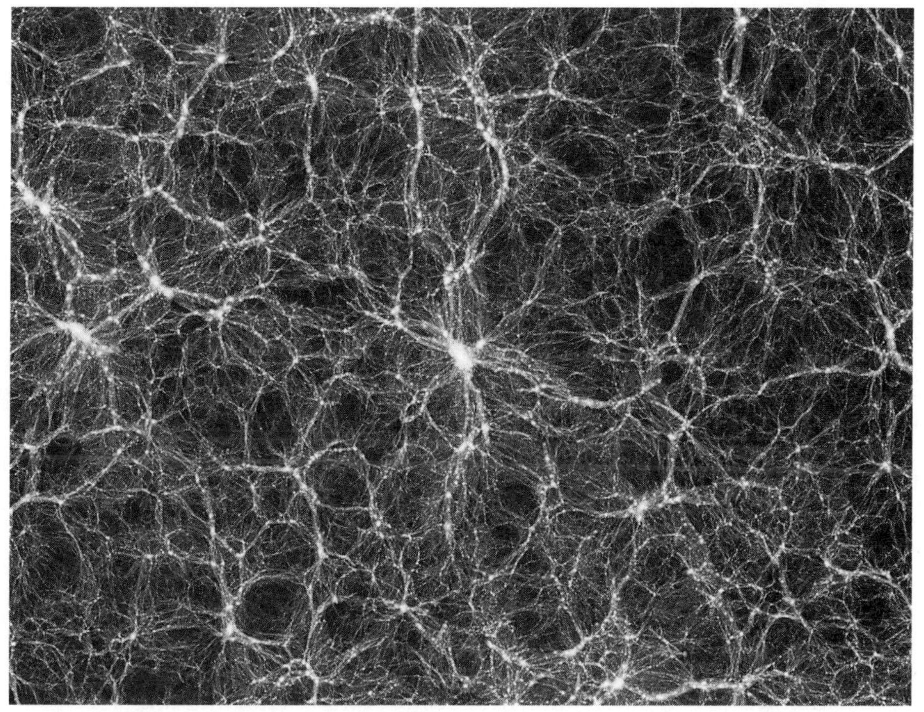

A slice of the universe from the Millennium Simulation. Shown here is the intricate web of dark matter that is the site of the formation of all observed galaxies in the universe. Image courtesy of Volker Springel and the Virgo Consortium.

ing of ideas—theory, observations, and numerical simulations. With rapid advances in technology and computing, we can now create high-resolution cosmological simulations, which have transcended their initial role in confirming observations and have started to drive science toward questions that define its frontier. This inversion happened as simulations became generative, providing a new and powerful methodology to create new knowledge rather than being just a limited tool to test ideas. They now offer insights into astrophysical processes that are not only extremely complex but also operate in cohesion in a way that simple pencil and paper calculations cannot predict.

The story of the journey to acceptance of the idea of dark matter has a completely different tenor than those of the two other radical ideas—the expanding universe and black holes—that we explored in previous chapters. First, the initial proposals for dark matter were completely empirically driven, and the theoretical framework to explain it was developed ex post facto. Second, there is no doubt that the invention of instruments and computational hardware and software was crucial to the discovery of the key role of dark matter in the universe. Third, what is salient about the process here is the many times that dark matter was discovered, discarded, and rediscovered before the idea ultimately won acceptance. It took Rubin and Ford's painstaking work documenting the speeds of stars galaxy by galaxy in a large sample and the simultaneous development of the dark matter theory for the idea of dark matter to be taken seriously. It is this confluence of theory—the development of the entire framework of the cold dark matter model of cosmic structure formation—and observations that led to the ultimate acceptance of the idea of dark matter. In addition to creating a new cadre of experts—simulators—the dark matter problem has underscored and refined the role of models in cosmology. The notion of a model that serves as a powerful intermediary between observations and theory rose to prominence when dark matter's significant role in the universe was recognized. Discovery of the dark matter particle, of course, is the big open question. Finding the body will solve the crime. Meanwhile, we cling to our belief in the tangible invisible and continue our quest for this mysterious particle.

5

THE CHANGING SCALE

The Accelerating Universe

In H. G. Wells's 1901 novel *The First Men in the Moon,* Arnold Bedford, a troubled London businessman, retires to the country to write a play that he hopes will improve his dwindling fortunes. As you may have guessed from the title, Bedford opts instead for the next best thing: embarking on a journey to the moon. His physicist friend Joseph Cavor's invention of a new material called Cavorite—a paste that allows anything coated with it to "deflect gravity"—makes the trip possible. After traveling in a spherical, Cavorite-coated spaceship, Bedford and Cavor land successfully on the moon. But only Bedford returns home safely, leaving Cavor languishing on the lunar surface, a captive of moon denizens called the Selenites.[1]

Wells's story highlights the perpetual human fascination with one of the central laws of physics. But the fantasy of defying gravity in the twentieth century was not just the stuff of fiction. In his 1948 essay "Gravity—Our Enemy Number One," Roger Babson, an MIT-trained engineer-cum-entrepreneur, reveals the origin of his wish to overcome gravity: the traumatic drowning death of his sister in her childhood and the tragic loss of his grandson, also by drowning. He explains, "Gradually I found that 'old man Gravity' is not only directly responsible for millions of deaths each year, but also for millions of accidents. . . . Broken hips and other broken bones as well as numerous circulatory, intestinal and other internal troubles are directly due to the people's inability to counteract Gravity at a critical moment." Babson, who had made his fortune by inventing new ways to statistically analyze stocks, founded two institutions in 1948: his eponymous business and entrepreneurship school, now called Babson College, and the Gravity Research Foundation, to battle his "enemy number one." Of course, any venture focused on defying a fundamental physical force is destined for failure; the Gravity Research Foundation folded on Babson's death in the 1960s but later reinvented itself by offering annual prizes to bona fide scientists for essays on the frontier of research into gravity. The foundation has since awarded many eminent scientists, including Stephen Hawking.[2]

Although Babson's crankish idea was a stretch even for science fiction, there is indeed a mysterious counterforce to the attractive pull of gravity. The closest thing to Wells's Cavorite might in fact be what we now call dark energy. Discovered in 1998, this mysterious force pervades the cosmos, driving our universe's accelerating expansion. As with dark matter, we have measured its effects but can't determine its fundamental nature. More troubling still, dark energy appears to be the dominant ingredient of our universe today. Evidence pointing to its existence has slowly accumulated since the 1980s, but only the direct measurement of the acceleration of the universe, and the discovery of deviations from Edwin Hubble's tidy law, confirmed it.

It was traumatic enough to accept Hubble's 1920s evidence of an unmoored universe. This added wrinkle—that the universe is not expanding steadily but rather accelerating at a breakneck pace—has proved even more discombobulating. Unlike the other ideas that this book explores, dark energy lacks an explanatory theory—and we seem to be far from that goal. As the science writer Richard Panek points out, the addition of *dark* before *matter* and *energy* "could go down in history as the ultimate semantic surrender. This is not 'dark' as in distant or invisible. This is not 'dark' as in black holes or deep space. This is 'dark' as in unknown for now, and possibly forever."[3] Our current understanding is merely a placeholder, and it remains to be seen if we will ever be able to unravel the nature of dark energy.

This isn't for lack of trying: the mystery of dark energy begins even before Wells's fictitious lunar landing. The theologian Richard Bentley questioned Isaac Newton in 1692 on the fundamental puzzle concerning the nature of gravity and equilibrium in the universe: if the universe is full of matter, and matter attracts other matter via gravity, why isn't the universe headed toward collapse? The question came as Bentley was preparing to give the first set of Robert Boyle Lectures in London, an honor that he likely owed to Newton's political maneuvering. Boyle, now famous for his work on gases and their properties, was an ardent believer and set up an endowment for lectures promoting religion. Bentley assiduously studied Newton's work and posed the question to Newton while working on preparing his lectures. In response, Newton acknowledged that his arguments required "that all the particles in an infinite space should be so accurately poised one among another as to stand still in a perfect equilibrium. For I reckon this as hard as to make not one needle only but an infinite number of them (so many as there are particles in an infinite space) stand accurately poised upon their points."[4]

Bentley became the master of Trinity College, Cambridge—Newton's alma mater—in 1700. As arrogant as he was ambitious, he had a controversial thirty-year tenure, and the fellows rebelled against his authoritarian style. Still, despite several attempts, they were unable to unseat him. Newton was staunchly supported by Bentley, who was incidently also shepherding the passage of the second edition of the *Principia* as Cambridge University Press was under his purview as master. Their friendship predated Bentley's reign at Trinity, having begun when he sought Newton's guidance in using the theory of gravitation to demonstrate that a divine agent must have designed the solar system. Bentley was looking for evidence of the divine hand to explain the physical interactions of material bodies that gravity entailed. His question to Newton was a profound physical one too. It probed right down to the heart of the matter—the principle of action at a distance that gravity engenders, which Newton had argued in favor of in a brief statement in the first edition of the *Principia,* in 1687. As a result of his subsequent detailed correspondence with Bentley, Newton replaced that statement with a more detailed treatment in a new section of the text, "General Scholium," that he added to the second edition, published in 1713. There Newton invokes divine action, explaining, "And so that the systems of the fixed stars will not fall upon one another as a result of their gravity, [God] has placed them at immense distances from one another." Newton appealed to a deity that would keep the planets stable in the solar system because he believed that mere chance would not have produced their configuration, nor kept it fixed over an extended period. Responding to Bentley, he asserted that "gravity must be caused by an agent acting constantly according to certain laws: but whether this agent be material or immaterial I have left open to the consideration of my readers." This was of course persuasive material with which Bentley could defend the existence of a divine order as per his remit for the Boyle Lectures. But despite the explanatory force of God, Bentley's question remained open, because of limitations in the understanding of gravity.[5]

If matters then required divine influence to hold a stable universe from collapsing, imagine the difficulty that explaining the expanding universe revealed in Hubble's time. If the universe were expanding at a fixed rate, then the linear relation between distance and velocity that Hubble found ought to hold as far out as we can see. In a universe filled with matter, however, expansion could never be constant, as matter would pile up in some regions and cause deficits in others. This in turn would drive collapse in the regions with mass aggregations and expansion in diluted regions. At the farthest distances from us, galaxies would have to deviate from the straight line of the Hubble diagram. Ascertaining if the universe's expansion is constant required measuring the Hubble relation out to the most distant reaches. To map and probe well beyond where Hubble and Humason did on the velocity axis of the Hubble diagram, astronomers could again use the redshift as inferred from the spectral lines of distant galaxies. The problem was how to measure distances accurately beyond the point where individual Cepheid stars could not be identified and tracked, as Hubble had done. Distant galaxies needed a more reliable, if not better, candle than the nearby Cepheids.

Hubble's contemporary and competitor Fritz Zwicky (of dark matter fame) made the key discovery that enabled the extension of Hubble's law to the farthest reaches of the universe. Zwicky identified a new class of bright beacons in the universe that are observable beyond where Hubble and Milton Humason had probed. In 1943, Zwicky and Walter Baade calculated that under specific circumstances, the core of a star can undergo a chain of nuclear reactions and collapse. The discovery by James Chadwick of the uncharged subatomic particle, the neutron, a decade before suggested that the implosion of a star would leave behind an ultracompact core of neutrons. This would happen after the violent expulsion of all the outer layers of the star because of shock waves generated during the implosion. The neutrons in the core left behind would be packed extremely densely. For example, one teaspoon's worth of neutron star would weigh about ten trillion kilograms! Following in Subrahmanyan Chandrasekhar's footsteps, Zwicky and Baade noted that the dying gasps of these stars would be extremely bright. They coined the name supernovae for them. Having predicted this end state of massive stars, Zwicky commenced an observational search for these extremely bright supernovae explosions. He even custom-designed an eighteen-inch telescope on Mount Palomar to hunt for these cosmic beacons. He did eventually uncover these end products of stellar deaths, but it was hard going and their tally grew slowly. His quest for supernovae and his discoveries made it into the *New York Times,* including a piece on September 27, 1942, "Another Exploding Star Appears in the Heavens," in the Science News in Review section. Previously, an English amateur astronomer had inadvertently discovered the first-identified supernova, which exploded twelve hundred years ago in the constellation Hercules. In a December 29, 1934, *Science News Letter* report, the astronomer Harlow Shapley had said that the discovery of supernovae might be the most important astronomical event of the time. Noting the similarities among the objects produced in these explosions, Baade suggested that they might indeed be standard candles in a paper he published in 1938, where he also cautioned that it might be a number of years before there was better data to establish them as such.[6]

Einstein too had a role in this game, yet again. You may have noticed that he is in the background—if not on center stage—in nearly every cosmological development of the past hundred years. It's helpful, then, to recall the story of how, in order to hold the universe steady, he tampered with his field equations and introduced the cosmological constant term, lambda, into his general theory of relativity. Recall as well that Hubble's discovery of the expanding universe rendered Einstein's constant incorrect, finally driving him to concede and accept a universe in motion.

As it turns out, the cosmological term that Einstein used to doctor the equations was wrong for then, but lambda appears surprisingly right for now—even if its job is much the opposite of what Einstein intended. In his field equations, this term represents a repulsive force that perfectly balanced the attraction of gravity to keep the universe static. Yet Einstein failed to appreciate that the cosmological balance that this creates is extremely precarious. Any small deviation from the status quo would drive the universe completely away from this perfectly poised state. It's rather like the feeling we get when we are standing up on our tippy-toes. One slight push, and we could just fall over.

But for the universe, a small nudge would mean more than a stumble. A slight change in the value of lambda in one direction would cause the universe to fly off into expansion—accelerated expansion, in fact—and likewise, a small squeeze would cause all existence to tumble (as Bentley imagined long before) into total collapse. Though Einstein seemed unbothered by this delicate equilibrium, Arthur Eddington understood its consequences.

Eddington paid attention to Vesto Slipher's velocity measurements and, even as early as 1923, began to ponder the deeper meaning of lambda. When Hubble was starting to see hints of the expanding universe, Eddington thought that this could point to a more nuanced role for Einstein's cosmological constant. In a public talk at the International Astronomical Union's meeting in Cambridge, Massachusetts, in September 1932, he not only promoted Georges Lemaître's solution of expansion from a big bang but also speculated on the possibility of a nonzero cosmological constant. Eddington saw his role in this quest as that of a cosmic sleuth: "I am a detective in search of a criminal—the cosmological constant. I know he exists, but I do not know his appearance; for instance I do not know if he is a little man or a tall man. . . . The first move was to search for footprints at the scene of the crime. The search has revealed footprints, or what look like footprints—the recession of the spiral nebulae."[7]

Unlike Einstein, Eddington viewed the lambda term as the solution and not the problem, and he anticipated that this force could power an acceleration beyond just the expansion that Hubble was measuring. He imagined that such an effect might well show up if the Hubble diagram could be extended to include data of galaxies at much larger distances. Note that acceleration is the rate of change of velocity, so to discern it, we need to look back further in time and farther out in the universe. At the time, this was hard to tackle observationally, because it was very challenging to measure distances out into the cosmos beyond where Cepheids, the only available cosmic yardsticks, could be seen.

As we have seen in his refusal to accept black holes, Eddington was a brilliant scientist prone to emotional investment in his ideas. Even when the rest of the astronomical community, including Einstein, had finally bought into the idea of an expanding universe with no need for lambda, he refused to yield. A 1932 paper that Einstein coauthored with Willem de Sitter dropped the lambda term but left the case open for later reexamination, noting, "An increase in the precision of data derived from observations will enable us in the future to fix its sign and determine its value."[8] Questions remained, however, even in the face of this new lambda-free consensus. Hubble's discovery demonstrated that the expanding universe was outracing the effects of gravity at that time. But was this always the case? Would it hold far into the future? And was it true in the distant past?

To understand the answers, we need to take a closer look at Einstein's field equations. His insight that underpins the general theory of relativity is how the shape of space, the contents of the universe, and the fate of the cosmos are interlinked. His field equations therefore precisely encode the mutual dependence of this triad—the shape, content, and fate of the universe. This triad also predicts many other properties, including the age of the universe. By the 1990s, astronomers had surveyed the contents of the universe, but the accounting was inconsistent with its shape and age—this left them totally stumped. General relativity was firmly established; its predictions had been observationally verified, so this mismatch of triad values had to be signaling something else entirely—perhaps a new missing ingredient? The geometry and evolution of the universe were puzzlingly incompatible with its contents. Starting with Vera Rubin and Kent Ford's findings, observations showed clearly that most of the material in the universe, matter, was exotic dark matter. Yet there wasn't enough of it to explain the inferred overall shape of space, and this dearth implied an age for the universe that was much younger than previously thought. The oldest stars in our galaxy seemed to predate the universe—clearly a problem.

When I refer to the shape of space, I mean not the local potholes in the otherwise taut fabric of space-time but rather the overall shape of the universe on the largest scales, where it appears more or less homogenous. Despite the growing wealth of data, including the measurements of the cosmic microwave background radiation (CMBR) from the Cosmic Background Explorer (COBE) satellite in the 1980s, there appeared to be a gnawing gap among measurements of the shape, the age, the contents, and the evolution of the universe. This intriguing lack of consistency propelled astronomers to measure these quantities to higher precisions. There was much to uncover. And more-accurate measurements were the need of the hour.

We know that lambda cannot completely resolve this inconsistency, because it cannot hold the universe steady in a stable and sustained way. (Remember, it puts us on tippy-toes—a solution ready for a stumble.) But if we dispense with the cosmological constant and stick with Lemaître's solution of an expanding universe, the expansion is described entirely as the result of the tussle between motion and gravity, specified by the average density of the universe. The observed density of all the constituents of the universe can thus be compared with a critical value—a tipping point that separates possible fates for the universe.

The ratio of the density of all matter and energy in the universe to the critical density is a pure number represented by the capital Greek letter *omega*. The total omega for the universe is the sum of the ratios of ordinary matter, dark matter, energy in relic radiation from the big bang, and the cosmological constant term to the critical density. The contribution to this budget from relic radiation today is negligible. So omega is really down to the contribution from matter and the contribution from the lambda term. The three potential solutions to Einstein's equations, which uniquely relate the universe's geometry, contents, and fate, correspond to three different values for omega.

First, let's consider a simple case, where there is just enough matter to not matter—that is, omega is significantly less than 1. In this case, the universe will expand at a constant clip forever, without slowing down or speeding up—a coasting universe. In this scenario, starting with a big bang, Hubble's law would hold true for every observer anywhere and everywhere in the universe.

If the universe instead had a considerable chunk of mass, say an omega with basically any value less than the tipping point 1 but greater than that of the coasting universe, then it would continue to expand, as in the previous case. Such a universe would get progressively diluted but keep slowing down as it expands.

Both of these scenarios are theoretically possible. The obvious way to determine which solution matches our universe best would be to measure omega directly. Looking at it piece by piece, we note that one way to estimate the matter density of the universe in order to measure omega is to add up the masses of all galaxies in a patch of sky, divide that by the volume they occupy, and put the result into a ratio with the critical density. When, in the 1980s and 1990s, astronomers did this by drawing on data from many galaxy surveys, they found that the matter density omega of the universe is about 0.3.

But there is another approach to independently measure the total omega. Because the CMBR—the relic radiation from the big bang—reaches us today after having rippled through the entire universe, it carries the imprint of the entire contents of the universe and thus allows us to estimate this sum as well. Measurements of the fluctuations in the CMBR detected by COBE in 1990 gave the answer that omega is equal to 1. Here was the puzzle: what accounts for the missing 0.7 in omega? The 0.3 measurement of omega that includes the contribution from all (both visible and dark) matter suggests that the universe is perpetually expanding, with a slight slowdown—a deceleration.[9] Astronomers began to search for signs of this slowdown.

It is important to stress that these two conflicting values for omega meant that something was seriously amiss. Do we live in a subcritical universe? Or one that was just at the tipping point? Remember Eddington's solution: there was just one surefire way to check which measurement was right—by determining if the Hubble diagram was valid much, much farther out. Since the expansion velocity is tied to the contents of the universe per Einstein's theory of general relativity, an observed change of the expansion velocity with time would point to the correct value for omega. To extend the Hubble diagram, astronomers needed new, extremely luminous beacons, because Leavitt's Cepheid variables—our first cosmological yardsticks—were simply too faint. It was time to search for a new cosmic standard candle, one that would be visible out to the edge of the universe. Supernovae, stellar explosions one hundred thousand times as bright as a Cepheid, were the answer. These intrinsically much brighter objects could be detected out to much larger distances. And because light travels at a finite speed, seeing farther out is equivalent to looking further back in time. If we inhabit a low-omega universe, then probing with supernovae might show that there was a slowdown of the expansion in the past—a deceleration.

If the only issue to settle was whether omega is 0.3 or 1, then it would simply be an accounting problem—find the missing matter. But there is a twist. If the cosmological constant lambda, abandoned by Einstein, is nonzero, then it also contributes to the estimate of omega. As you can probably guess, including lambda in the mix allows us to resolve these two inconsistent measures of omega in one fell swoop.

So while the cosmological constant is a bit of a nuisance, it also offers a possible handy way to reconcile the COBE estimate of omega with that inferred by measuring the density of galaxies. If the value of the cosmological constant is 0.7, then everything fits. But this would imply a slightly different and more curious fate for the universe. A tipping point universe with an omega equal to 1 is a very special case indeed. It corresponds to a flat geometry on the largest scales, with all the divots in space-time ironed out. So far we have focused on measuring the contents of the universe, but attempts to independently measure its geometry predated the efforts to perform a detailed inventory of its mass budget. Astronomers can also use standard candles to measure the geometry of the universe, whether flat or curved. Allan Sandage, Hubble's protégé, published a paper in 1961 outlining just such an observational program, to measure the geometry of the universe by detecting the present expansion and expected slowdown rate. This project was entirely in the context of a universe model with no cosmological constant and matter density below the critical value of omega equal to 1. To cover his bases, Sandage noted that if the cosmological constant were nonzero, it would imply an accelerating expansion for the universe rather than deceleration.[10] This curious flip of reality—acceleration versus deceleration—mentioned as an afterthought in Sandage's 1961 paper, was left unexplored for the next thirty-five years. Little did he know it would prove to be the stuff of Nobel Prizes.

Supernovae proved key to finding the answer. They were now in the astronomy toolkit as potential cosmological probes. But researchers needed to better understand their detailed physics to assess their feasibility as standard candles. All of this finally started to fall into place in 1985 when Wallace Sargent, an observer at Caltech, and his former graduate student Alex Filippenko, then a postdoctoral fellow at Berkeley, noticed that there were similarities in the spectra of many supernovae, suggesting that they were a class with some rather uniform properties. The variation of the apparent magnitudes of type Ia supernovae immediately following their explosion—the light curve—is remarkably uniform. And when these supernovae are glowing at their brightest, their spectrum reveals the fingerprint of the chemical element silicon in the explosion. This is just what was needed to establish supernovae as standard candles: the type Ia supernovae are extremely bright objects with "standard" similarities in their spectra that can therefore be used to peer farther into the universe and further back in time.

As with all astronomical objects, the apparent brightness of a supernova falls off as the inverse square of their distance from us. In the local universe, the redshift derived for a supernova from its spectrum is proportional to its distance (this follows from Hubble's law). Therefore, if we graph supernovae brightness versus redshift and if supernovae are standard candles, then these data should trace out and fall around a line. This is of course under the assumption that there are no changes in the expansion rate of the universe. But as we saw earlier, there could have been a change in the expansion rate during the travel time for light from the supernova to reach us. Essentially, if the expansion of the universe is decelerating (as expected for a low-omega-value universe with no cosmological constant), then a distant supernova will seem brighter than it would if the universe were expanding at a constant rate, as its distance estimate will be different. Likewise, if the expansion of the universe is accelerating —powered by a nonzero value of the cosmological constant, as Eddington noted—then the supernova will appear to be fainter at the same redshift than it would in a universe with a zero lambda. All of this hinges on whether supernovae are really standard candles—that is, whether they intrinsically all have the same luminosity. Although astronomers noticed similarities among supernovae explosions, there was still a bit of variation. To use supernovae as rulers or standard candles, astronomers needed to understand their physics better and to figure out how the small variations would impact their use as standard light bulbs. The clever solution was to calibrate these variations by studying nearby supernovae in detail and then to apply any corrections to "standardize" the distant supernovae that were in the same class, thereby pushing the Hubble diagram to the largest possible distances.

This was the idea that one of the pioneers in this field, Robert Kirshner, set off to pursue with his team of students and postdoctoral research fellows at Harvard University in the 1990s. He was on one of the teams that discovered dark energy in 1998.[11] Kirshner, a seasoned observer known for his cautious and thorough approach to research, was the enthusiastic and devoted mentor of Adam Riess and Brian Schmidt, whom we will hear more about later. By the late 1980s and early 1990s it was clear that to settle the question of the fleeing universe and reconcile its motion with its contents, chasing supernovae was the best bet. But they were the optimum measuring rods for the universe only if they could be found at large distances and carefully calibrated. Two independent teams of scientists, on opposite coasts of the United States, took up this challenge.

The Supernova Cosmology Project, based at the University of California, Berkeley, started on the race when it won a National Science Foundation (NSF) competition to establish new interdisciplinary research centers. Berkeley's Center for Particle Astrophysics proposed the marriage of two disciplines within physics with vastly different cultures—particle physics and astrophysics. It was the union of the study of the truly microscopic with that of the truly macroscopic. And the study of cosmic contents was one of the center's key goals. The other group, the High-z Supernova Search

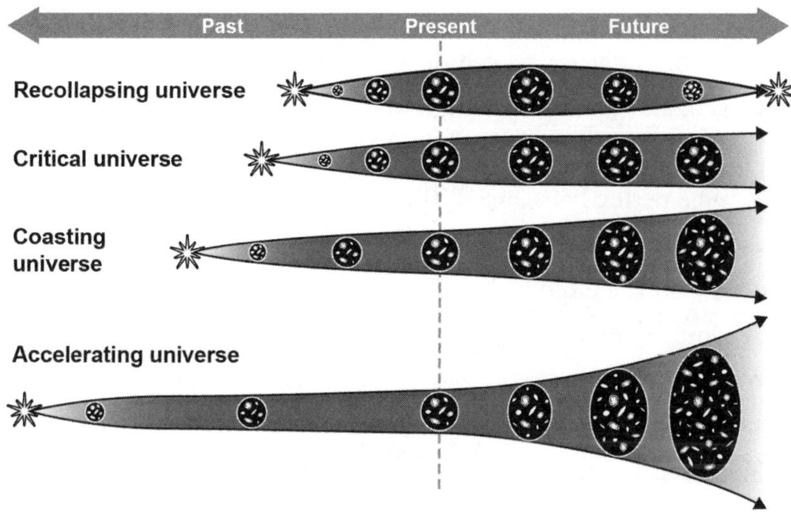

The fates of the universe in various cosmological models, showing how future and past behavior depend entirely on the values of cosmological parameter omega.

Team, with the largest number of members at the Harvard Smithsonian Center for Astrophysics, was composed primarily of astronomers, experienced observers who had honed their expertise on a range of instruments and telescopes spread across the world in high and dry spots. With the big bang confirmed by the CMBR and other data, in the late 1980s the open cosmological questions centered around making a more careful inventory of the universe and its constituents to determine its fate. Alexander Friedmann's solutions for Einstein's field equations admit three classes of dynamic solutions (none of them is a static universe), and the challenge was to figure out which one best corresponds to the observations of our universe. Was there enough matter in the universe to slow and eventually halt its expansion, causing it to reach a maximum size and then begin to contract? In such a universe, space would be infinite and shaped like the outside of a sphere. Or was the matter content so sparse that the universe would continue expanding forever without ever turning back? Space would be infinite even in such a universe and shaped liked a saddle. Or does the universe contain just enough matter to slow down its expansion and eventually bring to it a halt? In this kind of universe, space would again be infinite, but flat. This just-so universe would have to be perfectly balanced. These three options are often referred to as the Big Crunch (too much matter), the Big Chill (not enough matter), and the Goldilocks Universe (just right).

From just one measurement of the rate of change of the expansion of the universe, astronomers could determine which solution best describes our universe and therefore, by proxy, also determine its matter content and shape. Prior to the 1980s, it became clear that there was dark matter in galaxies and most likely also smeared in the spaces between them. Subsequently, as previously discussed, the matter inventory appeared to contribute to an omega equal to 0.3. Could the cosmological constant, lambda, then be equal to 0.7? There was only way to settle the matter—get more data, extend Hubble's diagram, and check for any changes in the expansion rate of the universe in the past.

Looking for supernovae became one of the key agenda items for the Berkeley Center for Particle Astrophysics after it received NSF funding in 1988. The center's primary focus, though, was dark matter, but it also worked to determine the full inventory of matter in the universe. Using supernovae as standard candles, its researchers investigated which of the solutions to Einstein's equations would best describe the fate of our universe—the Big Crunch, the Big Chill, or Goldilocks. Supernovae searches were not new and had continued since Zwicky's time, but it was clear that a team would need a smart observing strategy to discover large batches of them strewn about in the universe and catch them as they were exploding and at their brightest. The theoretically calculated rate suggests that we can expect one supernova to go off every hundred years or so in a galaxy. The only way to witness a large number of supernovae, then, is to observe as many galaxies as possible. Berkeley team members Saul Perlmutter and Carl Pennypacker, who were not trained astronomers, imagined that this question might take no more than a couple of years to settle. They planned to make use of automated observing techniques that Stirling Colgate had developed at Los Alamos National Laboratory in the 1970s. Colgate, an heir to the toothpaste fortune and a colorful character, was a creative and distinguished nuclear physicist. In the mid-1970s, he set up a thirty-inch telescope in the New Mexico desert and programmed it to look at a different galaxy every three to ten seconds. Automated observing with telescopes was starting to become standard then, but he pioneered this combination of automated repetition and searches for transient supernovae. An astronomer in search of supernovae would need to compare images of the same galaxy several weeks apart to see if a new bright flash of light had emerged. The brightness of a supernova is so immense that it can easily outshine its entire galaxy. With a clever strategy but a tiny field of view with his automated telescope, Colgate was not spectacularly successful in hunting down supernovae.

What Perlmutter and others distilled from Colgate's experience was that the area of sky that needed to be searched and scanned to identify supernovae had to be vastly expanded to garner enough of them for detailed follow-up. After coming to Lawrence Berkeley National Laboratory as a graduate student, Perlmutter had stayed on as a postdoctoral fellow. On May 17, 1986, the Berkeley team, which now included Perlmutter, found its first supernova. At this point the team consisted of him, Pennypacker, Richard Muller, and several students. A bit too optimistic, they predicted a detection rate of one hundred supernovae per year. And the supernovae they found initially, all residing in our local universe, were too close to say anything about cosmological scales where deviations from Hubble's law were likely to manifest. But these were the baseline supernovae that were going to help define and hone their class as standard candles. The pace was slow and supernovae proved slippery, with just two more appearing in 1986 and 1987. The team applied for funding to mount a camera on a telescope in Australia to cover the Southern Sky, hoping that expanding the search area would increase the rate of detection. They were strongly motivated to find supernovae to map the fate of the universe. Because of their brightness, supernovae can be seen far away despite the apparent dimming due to distance. And in a tremendous stroke of cosmic luck, really nothing but sheer luck, their brightening and dimming occurs on the reasonable time scale of a few weeks, making them extremely convenient for humans to follow and track. This serendipitous timing is extremely rare in the universe, where most physical processes typically take about a million years or so! Supernovae do, however, have their complications—they are indeed "'rare,' 'rapid,' and 'random.'"[12]

With improving technology, more powerful telescopes, and larger, more sensitive detectors, astronomers have continually pushed out into space and back in time. This was the case again with the hunt for supernovae. The hope was to measure the expansion of the universe since the big bang and pin down the epoch of deceleration: the universe was believed to be slowing in its expansion rate, with gravity serving as a cosmological brake. Therefore, it came as a complete and puzzling surprise in 1998 when the two independent teams analyzing distant supernovae discovered that the universe behaved completely contrary to expectation—it was not only expanding but also accelerating rather than slowing down.

When massive stars blow up they produce supernovae of a different ilk—type IIs, which are not standard candles, because the properties of the exploding star entirely determine their brightness. The stars that produce the standard candles, type Ia supernovae, are tiny white dwarfs that are locked in a binary system and grab gas from their companion stars. What is standard for this class of supernovae is their similar peak brightness and the pattern in which the light fades after rising to this maximum following the explosion. The reason the same patch of sky needs to be surveyed periodically is to catch these explosions at their brightest and then also trace their fading light. The shape of the light curve and the brightness at its peak are both needed to calibrate and use these supernovae as standard candles.

The technique that revolutionized this enterprise, in contrast to Zwicky's attempts in the 1940s and 1950s and Colgate's in the 1970s, was digital image processing. With computer-based digital image–processing software, the ability to swiftly prepare and rapidly compare large, sweeping images of the night sky—sifting out the flashing supernovae, storing the data, following up, and computing light curves—became possible. This radically transformed supernova cosmology. We saw how photographic plates, permanent records of phenomena in the night sky, impacted cosmology, in particular Hubble's work. Likewise, it was the newly developing sophisticated tools of digital image processing that eventually helped push the frontier beyond where Hubble had reached. As computers became faster and could handle more complex algorithmic instructions, dedicated software for real-time processing of astronomical images made huge leaps.

By the early 1990s the race between the two teams was heating up. Both had competitive access to computing and telescope facilities. The Supernova Cosmology Project had been formally set up and was headed by Perlmutter at Berkeley. Its success was due to software that allowed computers to find supernovae automatically by subtracting images taken of the same patch of sky at intervals of a few weeks apart. By the mid-1990s, Perlmutter and his collaborators from Europe, South America, and Australia were finding batches of supernovae. After a decade or so of sluggish finds, this newfound success led to prompt follow-up on some of the largest telescopes in the world. Finally, supernovae were living up to their promise as standard light bulbs, ready to be exploited as effective distance indicators. Both the Supernova Cosmology Project and the High-z team had members spread across the globe, enabling privileged access to a host of telescopes for rapid follow-up. The High-z team was formally assembled in 1994 and led by Schmidt of the Mount Stromlo and Siding Spring Observatories in Australia. True to national stereotypes, Schmidt, a soft-spoken Australian with big sparkly eyes, was a relaxed personality, and the all-American, key team member Riess was the intense one—together they were a powerful, harmonious duo. By 1993, it was clear that type Ia supernovae varied in brightness and were not quite perfect standard candles. Since several High-z team members, including Kirshner and Filippenko, were acknowledged world experts in the study of supernovae, the team concentrated on understanding the detailed physics of these explosions. The slight variations were better understood by then—it had been shown that the brighter type Ia explosions fade slightly more slowly than the fainter ones. Mario Hamuy, an astronomer based in Chile, and Riess, then a graduate student at Harvard, both on the High-z team, worked out how to use the curves obtained by measuring the rise and fall of the light from their explosions to calibrate type Ia supernovae as standard candles. To do so, any dimming of the light by dust in the galaxy where the supernova explodes needs to be estimated. Hamuy and Riess devised a method to correct for such obscuration and to accurately determine the supernova's maximum brightness. Taking this dimming effect into account and correcting for it is crucial for the use of supernovae as cosmic rulers.

Both teams tracked large swaths of the sky several times a year—after a new moon, when the sky was pitch black, for the best contrast. These were the baseline set of images. Three weeks later they went back and reimaged the same parts of the sky to then compare and contrast and see if any supernovae had exploded in the meantime. Although type Ia supernovae are not that common, both teams were resolute in working on beating the problem down with statistics. Both increased the number of galaxies that they examined—to hundreds of thousands in every image. This rate of scanning meant a typical yield of about ten or so per image selected for detailed follow-up. Of course, once the candidates were pinned down, the groups needed to follow the rise and fall of their brightness to derive their light curves. This further tracking required time allocations on ground-based telescopes. Crucial spectral information was also needed, to check for sure that these were indeed type Ias and to determine their redshift and hence their distance. Both teams were locked in competitive, fervent proposal writing—requesting time on telescopes and arranging for observers to travel, process, and analyze their data sets. It was quite an assembly line, which required cooperation and teamwork. Both teams were fairly well-oiled machines after the first year or two of gearing up the effort to hunt down supernovae.

By 1997 both had enough supernovae to draw and extend Hubble's original diagram from 1929. In doing so, however, they found incredibly bizarre and puzzling results. They remained cautious in their announcements, going only so far as to claim that they both seemed to have accumulating evidence for a low-matter-content universe. At this point what really amped up the game was supernovae data from the Hubble Space Telescope. The exquisite resolution of this telescope, orbiting in space, means that its measurements of light are significantly more accurate than any data taken from the ground, where the earth's atmosphere blurs light. Although astronomers can account and correct for this blur, such corrections add to the overall error incurred in the measurement. The race here was for accuracy and precision, and to achieve them required meticulously characterizing all sources of error. Once both teams analyzed their first data from Hubble, they found that its two initial supernovae data points were not lining up with their earlier trend. So they waited patiently to get more data, and once they had a handful of supernovae—six, in fact—the Supernova Cosmology Project submitted its paper to *Nature,* in the first week of October 1997.

This paper stops just short of making the case for lambda, the cosmological constant. The abstract concludes, "When combined with previous measurements of nearer supernovae, these new measurements suggest that we may live in a low-mass-density universe."[13] The High-z team posted its independent results, which it made publicly available, with its Hubble data on October 13, 1997. The submission to *Nature* meant that the results of the Supernova Cosmology Project were embargoed until the completion of the peer review process. The High-z team bravely concluded that matter alone is insufficient to produce a flat universe. Finally the two groups were starting to converge on their results and claims for the fate of the universe. It appeared highly likely that the expansion of the universe would continue forever. Eagerly looking to detect deceleration of the universe, both teams were initially stumped to find acceleration instead. There was much internal discussion within both before the announcement of their respective results. The American Astronomical Society invited both teams to participate in a press conference where they could present and discuss their results. Meanwhile, in the few months until the end of 1997, team members from both collaborations gave talks at colloquia and seminars around the world, hinting at and sometimes mentioning the possibility of a nonzero value for the cosmological constant, lambda. Riess and the High-z team felt they were falling behind, as they had fewer supernovae, and they upped the ante by carefully reanalyzing their data set. Riess had gained a reputation as the dust expert in his collaboration. He added in the twenty-one nearby supernovae that he had found during his thesis research to develop his light curve calibration method. These data had not been published previously. Using these supernovae to an-

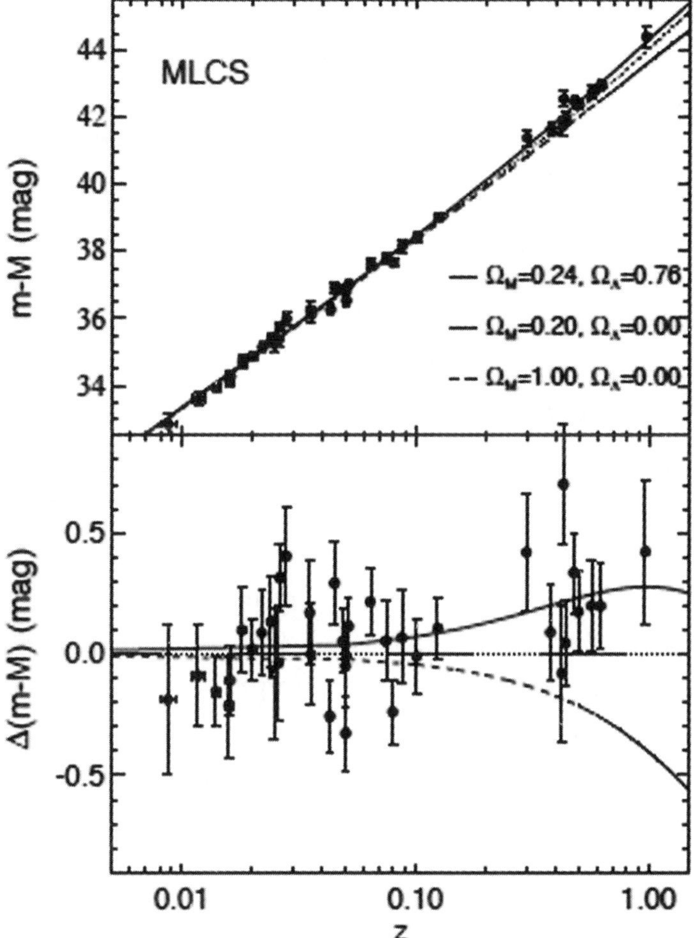

FIG. 4.—MLCS SNe Ia Hubble diagram. The upper panel shows the Hubble diagram for the low-redshift and high-redshift SNe Ia samples with distances measured from the MLCS method (Riess et al. 1995, 1996a; Appendix of this paper). Overplotted are three cosmologies: "low" and "high" Ω_M with $\Omega_\Lambda = 0$ and the best fit for a flat cosmology, $\Omega_M = 0.24$, $\Omega_\Lambda = 0.76$. The bottom panel shows the difference between data and models with $\Omega_M = 0.20$, $\Omega_\Lambda = 0$. The open symbol is SN 1997ck ($z = 0.97$), which lacks spectroscopic classification and a color measurement. The average difference between the data and the $\Omega_M = 0.20$, $\Omega_\Lambda = 0$ prediction is 0.25 mag.

Supernova Hubble diagram showing the Multi-wavelength Light Curve Shape method used to calibrate them, published in Adam G. Riess et al., "Observational Evidence from Supernovae for an Accelerating Universe and a Cosmological Constant," *Astronomical Journal* 116, no. 3 (1998): 1022.

chor the low-redshift part of the famous Hubble diagram, he now found that the solution that it was trending toward was a universe with not just no matter but negative matter![14] For concordance with other astronomical probes of the contents and geometry of the universe, say of the CMBR, there had to be something else added to the mix for omega to explain the puzzling supernova result.

This was alarming, and Riess started checking every bit of his analysis with his teammate Schmidt, who was doing an independent calculation with his own code. They developed a routine of emailing each other and then checking in on the results. Finally, on January 8, 1998, just ahead of the American Astronomical Society press conference, Schmidt emailed Riess, "Well Hello Lambda!" They had both arrived at the same result—they were staring at the evidence for Einstein's infamous cosmological constant, lambda—with the same degree of confidence, which they both estimated to be 99.7 percent. They reported this to the rest of the High-z team because a decision had to be made on how and when to share the results publicly. Given the momentous implications of this finding of a nonzero cosmological constant, several team members urged caution and warned that "press releases and a barrage of [*Astrophysical Journal*] *Letter*/*Nature* articles may impress the public or scientists who have only a casual interest in the subject, but the hard-core cosmology community is not going to accept these results unless . . . we can truly defend them." Schmidt, who had worked hard on the analysis thoroughly and independently with Riess, was confident of the result and reassured the collaboration in an email: "As uncomfortable as I am with a cosmological constant, I do not believe we should sit on our results until we can find a reason for them being wrong (that too is not a correct way to do science)." Filippenko, the Berkeley astronomer, who had defected from the Supernova Cosmology Project to the High-z team a few years prior, added, "This might be the right answer. And I would hate to see the other group publish it first."[15]

Perlmutter and his team at the Supernova Cosmology Project were working hard to prepare for the press conference too. The media lit up with the consensus result that both teams had arrived at on this most peculiar fate of the universe, evidently driven by a most mysterious agent—the cosmological constant. The Supernova Cosmology Project garnered more of the limelight initially because of its larger sample. The High-z team had presented only three supernovae. By the time of the annual American Astronomical Society meeting in January 1998, held in Washington DC, Perlmutter and his group were tentatively saying that they might be seeing evidence for the existence of a nonzero cosmological constant, which had originated as Einstein's alleged blunder and his fix to keep the universe static. That repulsive term, which would counterbalance the force of gravity, had made a comeback. Six weeks later, at a conference in Los Angeles, the High-z team, after correcting for a major uncertainty due to dust, also reported finding evidence for lambda. And this, they reported, was causing the universe to accelerate. They noted that the total density lay at a perfect balance of omega equal to 1. The CMBR and other cosmological observations had earlier hinted at an omega equal to 1. The realization finally dawned that this omega included ordinary matter, dark matter, and the cosmological constant. Soon thereafter, there was an article about "a universal repulsive force" in the journal *Science* by the science writer James Glanz, who had been assiduously following the project of finding and tracking supernovae.[16] At this point, both teams had only claimed evidence for a cosmological constant and had cautiously refrained from claiming a discovery. Declaring a discovery requires an extremely high degree of confidence in the methodology, analysis, and estimate of errors.

Finally, on February 22, 1998, Perlmutter presented the Supernova Cosmology Project's results at the University of California, Los Angeles, Third International Symposium on Sources and Detection of Dark Matter in the Universe, in Marina del Rey. Filippenko from the High-z team spoke right after. He walked up to the podium, paused, and then boldly declared, "Either you had a result or you didn't." The High-z team, he said, had one. Now it was public; the High-z team had a high degree of confidence in a cosmological constant that it was willing to defend, and solid evidence in hand. Supernovae that were distant a billion light-years or more appeared dimmer than expected, as the universe had expanded more rapidly since these stars had exploded, pushing them farther away from earth. And this accelerating expansion was powered by dark energy, manifested as the cosmological constant. Dark energy finally became real. Predictably, right after this, the tussle to apportion credit, both within and among the teams, began in earnest.[17]

Given this extraordinary finding, there were many questions about other possible explanations. Could it be that nature was playing a perverse trick—could older supernovae, which are farther away, simply be different beasts? The earliest galaxies were not as rich chemically as ones that assembled later—perhaps that caused their supernovae to be fainter? Yet the spectra of the supernovae both distant and near are similar; if there were a fundamental difference in composition, that would show up in the spectra. Therefore, both teams concluded that the more likely explanation was that the universe had accelerated, leaving the supernovae farther behind since the explosion that produced them. Because two independent teams with largely independent supernovae candidates as data and completely separate analysis techniques had arrived at the same result, after they had carefully convinced themselves, it was easy for them to convince the rest of the cosmological community too. Unlike in the past, there were no holdouts or powerful individuals who disputed these claims. Although the evidence for a cosmological constant was a radical discovery, because it aligned with prior mounting indications of an omega equals 1 universe from other cosmological probes—of the CMBR and other astronomical surveys—it was accepted swiftly. What also helped to ease the acceptance of the idea was that the cosmological constant was a familiar concept because Einstein himself had introduced it decades earlier.

The cosmological constant was also appealing because it resolved many other persistent disputes that had plagued cosmology. One key matter it definitively settled was the controversy regarding the age of the universe—a discrepancy discussed earlier, with rocks and stars dated as older than the supposed age of the universe, a bit of an embarrassment for the big bang and cold dark matter models. With the discovery of dark energy, the expansion rate of the universe was boosted, and therefore its time line needed to take this into account—and doing so suddenly made the universe older. The existence of dark energy thus also resolves the issue of why the matter contribution to omega is small even though a host of other astronomical observations found omega equal to 1. The cosmic inventory had been missing one critical ingredient.

Although dark energy is a convenient placeholder that helps tie in several observed properties of the universe, it is just that, a placeholder. As with dark matter, we know dark energy exists, but we have no real clue to its origin or evolution. So we now have a well-supported cosmic inventory, but the nature of the bulk of what composes our universe remains elusive. We appear to be living in a universe that contains only 4 percent ordinary atoms (everything we know of on the periodic table), 23 percent dark matter, and 73 percent dark energy. We have yet to come up with a theoretical framework that describes how and when dark energy came into being. Is it a field called quintessence—a previously unknown fundamental force—as some physicists suggest? Does it change with time? Is it fixed? These questions remain unanswered.

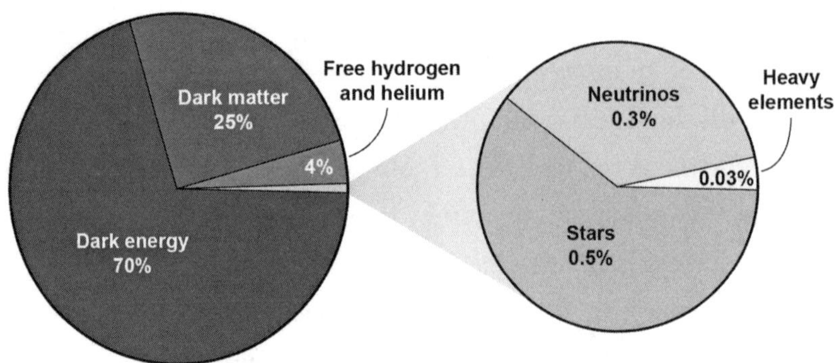

The pie chart on the left shows the total mass-energy sum of the universe, dominated by dark matter and dark energy, respectively. The callout on the right shows the composition of ordinary atoms in the universe.

How different are dark energy and dark matter from the mythical fluids—the ethers and effluvia—that were thought to exist in earlier times? We have empirical proof—observations, measurements, and many lines of independent evidence—that all point to the existence of dark matter and dark energy. We now have the instruments and technologies to probe their properties and are engaged in further investigation into their nature. There are many planned satellite missions and ground-based observational projects aimed at better understanding dark energy.

Rather unusually, the team leaders who led the observational efforts that discovered dark energy were awarded the Nobel Prize in 2009 for a discovery from 1998, a rather swift reward by normal standards in physics. According to Alfred Nobel's wishes, this annual physics prize can be awarded to no more than three people in a given year. This stricture has posed new problems for physics, and in particular cosmology, where, because of the intellectual maturity of the field, cutting-edge research is increasingly conducted in fairly large collaborations. The process of sifting out three individuals to honor for any pathbreaking discovery that was essentially a group effort is increasingly difficult, leading inevitably to unhappy researchers come the second week of October. Astronomy as a field is rarely recognized, but the 2009 award of the Nobel Prize in Physics went to Riess and Schmidt from the High-z team and Perlmutter from the Supernova Cosmology Project for the discovery of dark energy. This led to great debates about dark energy's many unsung heroes. Soon after the announcement, there were many public pleas for the Nobel committee to consider research teams for physics awards. No changes have been made to the rules, and the issue of partitioning credit and parceling awards is still fraught—in particular, because the accumulation of knowledge that leads to breakthroughs today has multiple key contributors whose collective work leads to progress. Some of the more recently instituted prestigious prizes have recognized this shift in the working culture of science and have begun awarding entire collaborations. The Gruber Foundation, for instance, has led the way, awarding its prize in cosmology in 2006 to the entire COBE team and in 2007 to both of the teams that discovered dark energy. In a practice pioneered by the COBE team, the discoverers of dark energy invited their entire collaborations to participate and share in the festivities in Stockholm on December 10, 2009. The Breakthrough Prize, founded by the Russian billionaire Yuri Milner in 2012, has been set up to award several individuals as well as teams if warranted. Milner, a former physicist who made his fortune in investment banking, is an enthusiastic supporter of the sciences in his philanthropy. He established Breakthrough Prizes in fundamental physics, mathematics, and life sciences. In 2013 the Breakthrough Prize in Fundamental Physics was awarded to the team leaders of the various groups that were involved in the discovery of the Higgs boson at CERN, and in 2015 it was awarded to both the Supernova Cosmology Project and the High-z Supernova Search Team. How to partition credit in a just fashion is one of the new challenges of the way that scientific work is presently organized and conducted.

The scientific questions that are at the forefront of cosmology today

—namely, unraveling the nature of dark matter and dark energy—require the efforts of large, international collaborations. As the historian of science Peter Galison writes, "Big science entails a change in the very nature of a life in science. Teamwork and hierarchy characterize daily work." This scale of operation originally characterized only giant accelerator projects like CERN in particle physics, but cosmology has transformed over the past thirty years into a field that is likewise no longer the domain of lone researchers. Instead, organized teams fight for funding and develop advanced methods and technologies to tackle and coordinate technically challenging tasks—in this case gathering and standardizing supernovae data. As the physicist Wolfgang K. H. Panofsky argues, the very kinds of questions that we have chosen to probe have necessitated this growth in scale: "We simply do not know how to obtain information on the most minute structure of matter (high-energy physics), on the grandest scale of the universe (astronomy and cosmology), or on statistically elusive results . . . without large efforts and large tools."[18] What is amply clear is that regardless of how the practice of science has transformed because of this scaling up of the enterprise, science cannot survive in isolation from other spheres of society, and its larger context has become ever more important. This is all the more so as scientific research now requires a huge allocation of resources human, technological, and financial.

The process of scientific acceptance of the more recent radical ideas in cosmology—dark matter, the CMBR, and now dark energy—has been somewhat different from the prior mode. The journey to consensus within the scientific community was swift and smooth for the surprising and intriguing observational discovery of dark energy, although it upended our view of the cosmos entirely. There are several reasons for the expeditiousness of this process. Bizarre though it was, the discovery of dark energy made many disparate observations fit together remarkably well. It was almost akin to the final piece in a jigsaw puzzle. For astronomers, the cosmological constant was and is just a number; for theoretical physicists, lambda was and is a deeper concept—the vacuum energy of the universe and a fundamental property of space. The measurements are beyond dispute, but the origin is still an open question. Another reason for this rapid consensus within the field is the intellectually mature stage that cosmology has reached and the concomitant transformation in the practice of this science. The current pace of discoveries is frenetic, driven by rapidly evolving innovative technologies.

How science is done has fundamentally changed in the past thirty years. Because collaborative international teams are conducting the research, the discoveries have also changed consequentially. There is no longer swift overthrowing (or even slow consolidation and honing) of an old theory but instead a deluge of data that raises as many new theoretical questions as it answers. The discovery of dark energy has challenged our deep understanding of physics and highlighted the gaps in our knowledge of the very early universe. Cosmology has also led the way in the ongoing big data revolution in all intellectual disciplines. A new frontier remains to be explored further—the true nature of dark energy—and this has opened up novel sets of questions of an unprecedented kind that hark back to the moment of creation.

6

THE NEXT WRINKLE

The Discovery of Cosmic Background Radiation

After a sleepless Saturday night, John Mather and other members of the Cosmic Background Explorer (COBE) team drove in the ghostly predawn light of November 19, 1989, to the space and missile testing facilities at Vandenberg Air Force Base near Santa Barbara, California. As Mather toured the viewing sites scattered around the launchpad, he was keenly aware of their mission's high stakes. He knew that a successful launch could radically transform our understanding of the universe: COBE was designed to measure the remnant hiss from the big bang. As this leftover radiation from when the universe was hot, dense, and about four hundred thousand years old races toward us, it traces the expansion history of the universe. On that California morning the air was thick with excitement, tension, and anticipation. Only hours before, engineers had to completely replace the Delta rocket launcher's onboard computer, and now the COBE team waited to watch the "satellite of love," which many of them had worked on for decades, rise and disappear into the dark morning sky.[1]

Just as the launch countdown began, weather balloons already sent aloft indicated strong winds. Vandenberg halted the launch. Many worried they would miss the brief thirty-minute window and would have to reschedule the launch. After what must have seemed like aeons to the assembled spectators, and much to their relief, the countdown resumed. The crowd saw a flash of light illuminate the sky and watched a slow, mesmerizing lift-off. The Delta rocket deployed like clockwork; within ten minutes of launch, the first stage fell off and the second stage fired. The COBE satellite, now ejected from its carrier, cruised into its orbit, 170 kilometers (105 miles) up and out in space.

But it was weeks before the scientists on the COBE team could really celebrate—only after confirmation that all the instruments performed as expected in the cold vacuum of space. "First light," the successful commissioning and gathering of the first set of data from COBE, began after careful calibrations and checks. Once the data started flowing steadily, it was simply astounding; it faithfully followed theoretical expectations with exquisite precision. It was months before scientists from the COBE team were ready to present this data to other scientists and the public after diligent scrutiny and analysis. But when they did, they revealed an entirely new map of the universe, one that agreed to a remarkable degree with the predictions of cold dark matter theory. This was a map that captured ancient light from the deepest recesses of the cosmos.

Preparation for the COBE mission required herculean effort, much like that for a cartographic transformation of yore. In early 1519, Ferdinand Magellan presented a bold proposal to the Spanish king Charles I. Magellan was convinced that he could discover a new commercial route to Asia. But trade benefits aside, he was driven by the spirit of adventure and the desire to explore far-flung corners of the earth. After staving off many obstacles, including attempted sabotage by the Portuguese crown, Magellan eventually garnered all the required funds, including a contribution from a private merchant. Finally, he set sail with his fleet on September 20, 1519. One imagines that the sentiment at Sanlúcar de Barrameda, downstream from Seville, on that day was similar to the thrill that the COBE team at Vandenberg experienced 470 years later. It takes the same kind of tenacity today to assemble a scientific team, gather funding, design and produce an experimental setup, successfully launch it into space, and then interpret the data streamed back to earth. Both of these voyages of discovery were propelled by the same essential, deep human impulse and ambition, to explore. And both were highly risky. Magellan succeeded in circumnavigating the globe and revamped our understanding of the world with that one voyage. It was the harbinger of a new era of world trade and globalization. He helped remake our terrestrial map. Mather and the COBE team reshaped our understanding of the cosmos. For their efforts, the team members Mather and George Smoot were jointly awarded the Nobel Prize for Physics in 2007.[2]

The real shift in our understanding of the universe came from COBE's measurement of the cosmic microwave background radiation, or CMBR. The idea of the CMBR was first proposed in the 1940s, but few immediately recognized its importance. The experimental tools that confirmed its existence were devised for entirely different purposes. In this case, military radar developed at the Massachusetts Institute of Technology Radiation Laboratory during World War II proved to be key. The military-industrial complex that was set up for the war effort catalyzed numerous unanticipated advancements in basic science. As Helge Kragh has pointed out, cosmology was a special beneficiary.[3] But before delving into that, let's first go back to the beginning, literally to the beginning of the universe itself.

In "L'expansion de l'espace," published in the November 1931 issue of the obscure journal *Revue des questions scientifiques,* Georges Lemaître wrote, "The atom-world was broken into fragments, each fragment into still smaller pieces. . . . The evolution of the world can be compared to a display of fireworks that just ended. . . . We can conceive of space beginning with the primeval atom and the beginning of space being marked by the beginning of time." Of course, the expanding universe model that he had proposed implies a fiery beginning—the big bang. Lemaître also published a letter in *Nature* in 1931 expounding his theory of the origin of the universe from the "primeval atom" on "a day without yesterday"—a very dense point of space and time.[4] He was convinced that scientists could obtain material evidence for this beginning, that observational access to the primordial universe was inevitable. But at this juncture, the conversation and collaboration between theory and observations had just started, spurred by the success of the symbiosis between Hubble's observations and Lemaître's theory in establishing the expansion of the universe. These were early days for the kind of synergy and cohesion that eventually put theory and observations in lockstep, as they are today. The coupling of Lemaitre's theory and Hubble's observations opened promising new doors for cosmological research. Before this, no one had had the audacity to model the entire universe and chronicle its evolution from the big bang forward. Further exploration of the big bang model and its potential observational consequences occupied many physicists from then on. The big bang model, though, was far from established—it had severe competition. In particular, from the late 1940s, the steady state theory of Fred Hoyle and company provided growing resistance. Given this tussle of ideas about the origin of the universe, there was enormous interest in obtaining observations that might help discriminate between them.

One of the protagonists in our tale is the Russian émigré physicist George Gamow, who taught at George Washington University in Washington DC in the late 1940s. Along with two young colleagues, Ralph Alpher and Robert Herman, he began investigating the creation of chemical elements in the universe implied by the big bang model. Gamow was convinced that finding an explanation for the production of known chemical elements in the periodic table would nail the big bang model. He was viewed as a creative genius and was a font of new ideas, which he generously shared with students and collaborators. Like Fritz Zwicky, however, Gamow had a complicated personal reputation. In addition to his towering scientific contributions, he was known for his pranks and his alcoholism, which diverted focus from his many achievements. Gamow had an impressive intellectual lineage; he had studied at Petrograd/Leningrad State University (now Saint Petersburg State University) under Alexander Friedmann, the famous physicist who had produced the evolving universe solutions to Einstein's field equations. Gamow's pioneering and acclaimed scientific work was in radioactivity and stellar evolution. Persisting after many unsuccessful attempts, he and his wife finally managed to defect to the West, ending up in the United States in 1934. In spite of his many original contributions to and expertise in radioactivity and nuclear fusion, Gamow was kept out of the elite national scientific project—the Manhattan Project. He was not granted security clearance, although he was later invited to work briefly on the hydrogen bomb program at Los Alamos National Laboratory. It is now speculated that the reason for this was his known alcoholism. Despite his patchy personal reputation among his colleagues, he had a huge fan base in the public sphere. By the 1940s he was very well known as a science popularizer and the inspiring scientist author of many best-selling books, including *One Two Three . . . Infinity*, *Mr. Tompkins Explores the Atom*, and *Mr. Tompkins in Paperback*.[5]

It was around 1944 that Gamow began working in earnest with Alpher and Herman on cosmic chemistry. Herman, a recent PhD from Princeton, had worked on both expanding the idea of Lemaître's primeval atom and the beginnings of the universe. Gamow wanted to investigate the grand question of how all the known chemical elements had been synthesized. In particular, he wanted to explore if it was possible for all elements to have been made in the early universe, well before the first stars came into existence. He believed strongly in the hot big bang model, a universe with a hot, dense beginning, and was deeply invested in finding ironclad proof, which was lacking at the time. Since it was known that hydrogen and helium could be made in the nascent universe, he surmised that finding a credible explanation for the early origin of the rest of the elements would be the clincher. Adopting a novel methodology, Alpher and Herman extrapolated to the initial conditions, when the density of the universe was very high, starting from the current situation, wherein the universe essentially contains mostly hydrogen- and helium-rich stars. The early universe they estimated would have been so dense that particles and their physical nemeses, antiparticles, could form and destroy one another continually, thus allowing energy and matter to morph into each other rapidly. At extremely high temperatures, Einstein's equivalence between mass and energy (expressed in the famous formula $E = mc^2$) is realized—so particles and antiparticles abound. If the early universe was indeed a soup of unruly particles, Alpher and Herman realized, then this constant mass-energy transmutation would likely bring some kind of balance, and as a result the now familiar subatomic particles—protons, electrons, neutrons, photons, and neutrinos—would then condense into being as the universe expanded and cooled.

This kind of balance is referred to as thermal equilibrium, and it has some unusual properties. Imagine a closed box with opaque walls that trap energy (all forms of radiation, including light) and matter inside. According to quantum mechanics, such a box will reach equilibrium and behave like an ideal "blackbody," so that the intensity of radiation it emits will depend only on the temperature of its walls. It was Gamow who first realized and discussed the important role that thermal radiation and thermal equilibrium might play in the synthesis of chemical elements. Following up on his hunch, Alpher and Herman had a key insight. They surmised that the hot, dense, early universe, on achieving thermal balance, would have behaved like a blackbody. And since the identifying feature of blackbodies is their temperature, Alpher and Herman estimated the resulting cosmic temperature—the current temperature of the universe. Furthermore, they conjectured that even though expansion had cooled the universe, an indelible signature of the early, heated universe must still survive in the form of blackbody radiation. This pervasiveness of the radiation would arise because of the special shape of emission from a blackbody; a blackbody remains one forever, even as it cools. Therefore the universe would still be a blackbody today, albeit with a lower temperature than at its fiery start. Alpher and Herman estimated the current blackbody temperature to be about five kelvins (–268 degrees Celsius or –450 degrees Fahrenheit). That the universe is a blackbody and that the early universe and the present-day universe have unique temperatures were remarkable claims. Alpher and Herman's predicted low value for the current cosmic temperature was counterintuitive yet famously close to the value that was measured decades later. High temperatures are somehow more graspable from everyday experience, as we witness boiling water and food sizzling on grills. But this cosmic temperature of five kelvins was way below these familiar temperatures and way below our body temperature, which is about 310 kelvins, and well below the temperature even of ice. Temperatures aside, the primary goal for Alpher and Herman's calculation was to explain the origin and the buildup of atoms from the primeval fireball. They succeeded only partially in this task—only a tiny amount of any element heavier than helium was produced, however hard they tried. They published these new yet unsatisfactory results in the journal *Nature* in 1948, with the estimate of the extremely low current temperature of the universe tucked away in an aside.[6] This paper provided major insights into the physics of the early universe and corrected errors that Gamow had made in one of his previous papers, but it failed to explain the absence of a stable isotope with atomic number 5 in nature. This was a key part of the problem that they had set out to solve. Despite this, Alpher and Herman had many interesting results, including the calculation of the evolution of the density of matter in an expanding universe. However, since they had failed to account for the missing atomic number 5 isotope, and had failed to properly account for the origin of all the heavier elements, the paper was seen as a dud. Unfortunately, as the estimate of the cosmic temperature that they had computed was nestled in a paper that was considered wanting and faulty, it was ignored.

There was another complicating matter: their predicted value for the temperature of the universe did not convince even Gamow, their close collaborator! In a valiant attempt to connect his work on early element formation with his previous work on stars, Gamow variously predicted the temperature of the universe and of interstellar matter to lie anywhere in the range from five to fifty kelvins (–223 degrees Celsius or –370 degrees Fahrenheit). These roving and unsteady predictions from Gamow made Alpher and Herman's temperature calculation go further unnoticed. Preoccupied with explaining the origin of the elements and getting it right, Gamow, Alpher, Herman, and their collaborators published a total of eleven papers on the subject in 1948, none of which solved the cosmochemistry problem. Despite this prolific output, the scientific community completely ignored Alpher and Herman's prediction. This unfortunate fact, though, makes this issue the perfect case study of why some radical ideas have a more tortured journey to acceptance and how nonscientific factors come to play a significant role in the process. This is a famous instance when the entire scientific community overlooked an important contribution for well over twenty years, and as a result, there have been several careful, scholarly investigations afterward to understand why this pioneering work failed to make an impact. While not all the details of Alpher and Herman's personal

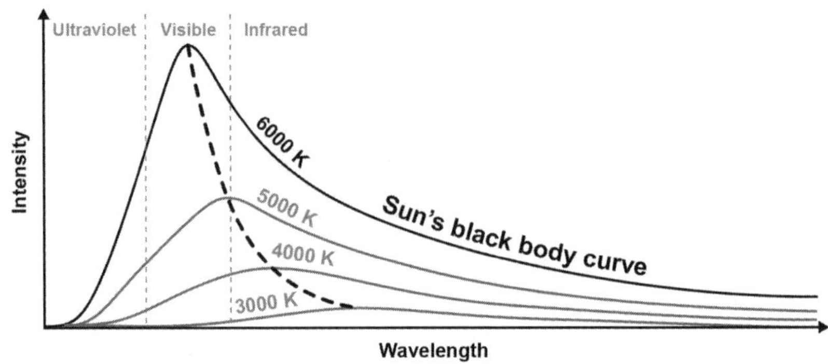

Sketch showing the shape of the blackbody spectrum of an emitting body in thermal equilibrium. Since the early universe behaves like a blackbody, the measured spectrum of the CMBR matches the curve perfectly.

motivations and intents can be accurately retraced ex post facto, the cosmologist James Peebles followed the trail of their scientific papers in a painstakingly researched article fittingly titled "Discovery of the Hot Big Bang: What Happened in 1948," and John Mather and John Boslough delve deeply into the matter in their book *The Very First Light*.[7] All the reconstructions of the history of this calculation agree on one thing—that Alpher and Herman were the first to estimate the temperature of the CMBR and to identify it as the temperature of the universe, in their 1948 *Nature* paper.

If we can speculate on what might have been a legitimate intellectual reason for this oversight, it was likely the fact that the paper's main subject and focus of calculation was not the cosmic temperature but rather the production of chemical elements in the early universe. The estimate of the temperature of the universe was an aside, so marginal that the paper does not explicitly address the issue of its potential detection or verification. Even if this prediction had been paid attention to, in 1948 it would have been a hard measurement to make, given the low value of the expected temperature. Any deliberate attempt to seek to measure the cosmic temperature came much later. In *The Very First Light,* Mather and Boslough imply that this measurement was technologically feasible but challenging in 1948. Strangely, Robert Dicke, a physicist at Princeton University, had sort of tried but did not realize that he had come close to measuring the temperature of the universe. His project involved the measurement of the blackbody temperatures of the sun and the moon. He noted that the thermal radiation from the atmosphere affected his measurements as a contaminant. Therefore it also needed to be measured. Dicke and three colleagues published a paper on this subject in 1946, just two years before the explosion of papers from Gamow, Alpher, and Herman; they concluded that the level of radiation implies that the temperature of the atmosphere is less than twenty kelvins (–253 degrees Celsius or –424 degrees Fahrenheit). They also estimated that the strength of this signal would be too weak for direct detection with a radiometer.[8] Dicke attempted this measurement in Florida, was unsuccessful, and soon lost interest. Since the significance of the idea of a cosmic temperature was unclear in the grand scheme of things, it was ignored for a good while. Besides, there was real confusion over whether the temperatures of the atmosphere and of interstellar matter were one and the same as the overall cosmic temperature. The lack of conceptual clarity certainly did not help. As we will see later, it is ironic that Dicke was the person who came close to making this measurement, with a device that he himself invented!

Part of the problem also stemmed from overspecialization and the lack of communication among scientific subfields in astronomy and physics. Herman and Alpher were stellar astrophysicists—a distinct subdiscipline—by training and thus perceived of as somewhat disconnected from cosmology. Helge Kragh, in his book *Cosmology and Controversy,* suggests that the conflicting and multiple values for the temperature of the universe (the CMBR) that Gamow, Herman, and Alpher published might have confused astronomers. Besides, it was unclear whether the temperature was of the cosmic origin or if it included the effects of stellar radiation. This confusion in interpretation was in fact an old one and likely led to the dismissal of what might have been truly the first reported detection of the CMBR, even prior to Dicke's aborted attempt. In 1941 the Canadian physicist Andrew McKellar had measured the temperature of the universe without realizing it. He was looking at the carbon-nitrogen cycle as the potential energy source for carbon stars. Thus the toxic organic compound cyanogen (which consists of carbon and nitrogen atoms) had sparked his interest, due to its detection in the tail of Halley's comet in 1910. While studying the spectra of molecules in interstellar space to understand their function and origin, McKellar determined that the poisonous cyanogen molecules were extremely cold—in fact, around three kelvins (–270 degrees Celsius or –454 degrees Fahrenheit). No one recognized the importance of this absurdly low temperature or that it had any meaning beyond McKellar's apparently esoteric study. The confusions about exactly what they were measuring aside, Kragh argues that Alpher and Herman made another crucial lapse by not discussing or even speculating on how the temperature that they had estimated might be measured. For example, their 1948 *Nature* paper fails to calculate and present the precise wavelength at which this cosmic temperature could be detected. In fact, it does not even have a plot of the blackbody curve. Kragh contends (and I agree) that omitting to mention that the cosmic temperature would be observable in the microwave band further deterred experimental efforts to detect it. Additionally, this lack of guidance to focus on the microwave band prevented astronomers from connecting the dots. All of Alpher and Herman's calculations appeared in the context of what was seen as an entirely separate field of study, nuclear physics. Cosmologists simply failed to take notice.[9]

As theorists, Alpher and Herman were looking to understand how chemical elements formed in the early universe, and this left them at a disadvantage. They weren't looking for the temperature of the early universe, and at some level they didn't believe anyone could find it. Meanwhile, they weren't successful in their stated quest either. What they found contradicted Gamow's initial hypothesis. Alpher and Herman realized that there was a window of only about the first few minutes in the early universe when hydrogen and helium could fuse to make heavier elements. Once this window was shut because of rapid cooling driven by the expansion, additional fusion (and thus the creation of heavier elements) would have been almost impossible. Alpher and Herman published this result, but one imagines that they felt rather demoralized at having failed to solve the fundamental problem they sought to answer.

There are several additional practical reasons why Alpher and Herman's important insights made no impact and faded into obscurity at the time. Because of the vagaries of the academic job market, both quit academia and moved to work in research labs at General Electric and General Motors, respectively. They stepped out of the field, publishing only occasionally, as and when they could snatch time from their day jobs to pursue their research. Their departure from the academic fold meant that they could not lead research teams into understanding the implications of their findings—the kind of work possible only with the help of armies of trained postdoctoral researchers and doctoral students, available only to scientists in secure, tenured positions at universities. Both Alpher and Herman had departed from the milieu where they could have had protégés working for them and disseminating their work, which, as we have seen before, helps establish intellectual legacies in science. Scientific ideas have always needed to be peddled and pushed by their instigators. Without the echo chamber provided by intellectual empires constructed through the recruitment of graduate students and postdoctoral researchers to elaborate and verify them, ideas dissipate and die. And crucially, the hot big bang theory still had doubters in the 1940s and 1950s. The CMBR is tied intricately to the big bang model and is in fact one of its key observational signatures. Since Alpher, Herman, and Gamow had not succeeded in explaining the origin of chemical elements in the context of the hot big bang model, this did not help persuade its detractors.

Another hurdle for full acceptance of the hot big bang model concerned estimates of the age of the universe. By looking at the rate of its expansion and estimating its size, astronomers can estimate its age. Measurements of the Hubble expansion provided an age for the universe of a mere two billion years. Meanwhile, geologists had already excavated rocks on earth that were several billion years older. This posed a conundrum. The hot big bang model seemed an incomplete, if not an inadequate, description of the universe. There was also the niggling issue of the beginning of the universe—why and how did its initial conditions arise?

At this juncture, in the late 1940s and early 1950s, the rival theory of the steady state universe, proposed by the Cambridge trio Hermann Bondi, Thomas Gold, and Fred Hoyle, had its philosophical attractions. Their conception was of an eternal universe. This allowed them to dispense with the need for an explanation for a beginning and thus forego the uncomfortable question of what came before the big bang. The steady state model gathered adherents, and many physicists were caught in the grip of its attractions. Alpher and Herman's prediction of the blackbody cosmic background radiation was incompatible with steady state cosmology, which is another likely reason that the CMBR lay forgotten until about 1965.

Another key factor for this very important idea getting lost was the intellectual climate of the time. The idea of a primordial explosion leading to an expanding universe was simply not yet established as serious science. Cosmology as a field was viewed with skepticism. Unlike many other branches of physics, it did not permit controlled experiments, and measurements on cosmic scales appeared to be plagued with unknown errors. It certainly did not have the feel of a traditional science. The intellectual space that cosmology occupied was due to change soon, though, and in fact, the accidental discovery of the CMBR catalyzed this process. And of course an idea, in particular a radical idea, often needs time to gestate, so that when propitious circumstances finally do coalesce, it can be swiftly evaluated and accepted if correct.[10]

Thus it was sheer serendipity that eventually catapulted our understanding of the early universe, and it came from a completely unexpected quarter. There is a saying in physics that one person's noise is another person's signal—what this means is that something that is effectively a nuisance in answering one scientific question often turns out to be extremely valuable and unexpectedly illuminating for a different one. This was literally the case for the first reported measurement of the CMBR, by two physicists who weren't even looking for it.

In 1964, two talented physicists at Bell Laboratories, Arno Penzias and Robert Wilson, were tinkering with the lab's horn antenna on Crawford Hill in Holmdel, New Jersey. Penzias, with a doctorate in experimental physics from Columbia University, had been on the job at Bell Labs for three years. Ambitious and sharp, he was good at grasping the big picture in any problem that he worked on. Wilson, a twenty-seven-year-old newly minted PhD from Caltech, had been schooled in the steady state theory and had just joined Bell Labs. He was a thorough type—a tinkerer at heart—fascinated with and a stickler for details, and he came with a reputation for technical prowess in getting instrumentation to work. For his doctoral thesis, he had mapped the Milky Way in the long, radio wavelengths. Little did he know that he would stumble into and prove instrumental in remaking our entire cosmic map.

Penzias and Wilson used the Crawford antenna as a telescope to study sources in the sky that emitted radiation in the radio spectrum rather than in the optical wavelengths. Radio frequencies, part of the electromagnetic spectrum, which includes visible light, range in wavelength from a few millimeters to about ten meters (thirty-three feet). They can penetrate and travel through earth's atmosphere without being reflected back or distorted. Although Heinrich Hertz discovered radio waves in 1887, it was Karl Jansky who detected the first radio emission from the center of our galaxy, in the 1930s, with a long antenna that he had built on the Holmdel site in New Jersey. The astronomy community—focused on optical light, lenses, and the spectrograph—did not immediately value this new window onto the universe. Yet after World War II, with the preeminence of radar (the acronym of "radio detection and ranging") and the concomitant large number of physicists and engineers who had developed it and were trained in its use, radio wavelengths saw a surge as a new tool for exploration in astronomy. This sparked interest in looking for even more cosmic radio sources.

The Crawford telescope that Penzias and Wilson worked with had a very sensitive radio receiver, assembled as a transponder and amplifier for the Echo communication system. Built in 1960 by Bell Labs, this system, which reflected radio signals off two large metallic balloons in the upper atmosphere, had enabled communication over large distances. By the time that Penzias and Wilson used the receiver, however, it was obsolete for communication, replaced by satellites. Antennas on radio telescopes needed to be able to pick up signals above the usual din to detect potential cosmic sources. The Crawford was tuned to a frequency of 4,080 megacycles per second, and this was perfect to detect signals in the microwave band. To find and isolate the extremely weak signals expected from astronomical sources at these wavelengths, Penzias and Wilson, looking for emissions from hitherto undiscovered radio sources, had to identify all the sources of static noise that could potentially drown their feeble signal, quantifying and therefore eliminating all sources of contamination.

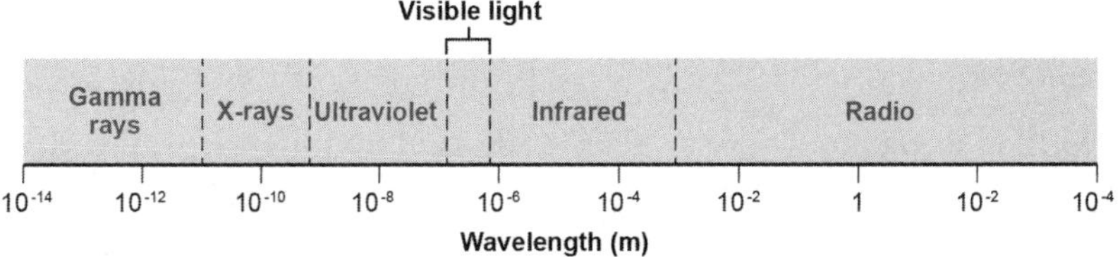

The electromagnetic spectrum, showing the full range, from gamma rays (shortest wavelength) to radio waves (longest wavelength), and clearly illustrating the narrowness of the visible band.

Despite their best attempts to identify and correct for all such sources, they failed. They were detecting static that appeared to drown their signal, and they wanted to remove it. So they checked noise from the ground, from molecules jittering in the upper atmosphere, from the workings of the apparatus itself, and even from two nesting pigeons that fancied sitting on the antenna. But the noise never subsided, regardless of the season or the direction of the antenna. They were stumped. This noise could not be identified and hence could not be removed. It was uniform both in space and in time, arriving from all over the sky, with no preferred direction. And the temperature equivalent of these feeble radio waves was roughly three kelvins.

Returning from an astronomy conference in December 1964, Penzias discussed this persistent noise issue with a radio astronomer colleague, Bernard Burke, who pointed him to a new paper by a young theorist named James Peebles, working under Robert Dicke at Princeton.[11] Dicke had become an established and extremely versatile physicist and was considered a world-class experimentalist as well as theorist, a reputation that he had developed while working on the war effort at the MIT Radiation Laboratory. Born in 1916, he was a soft-spoken, energetic man who was just as comfortable machining and designing electronic circuitry as solving complicated mathematical equations. He had come very close to inventing the laser. Charles Townes and his brother-in-law, Arthur Schawlow, filed their patent application for masers (the acronym of "microwave amplification by stimulated emission of radiation") in 1958; Dicke had filed to patent a very similar method of building the same kind of instrument, but one that operated at a different wavelength—an infrared laser—two years prior, in 1956. Townes and Schawlow's invention was a more comprehensive device, which could operate on a broader swath of the electromagnetic spectrum than Dicke's, which was restricted to infrared wavelengths. Although Dicke was awarded the patent in 1958, much of the glory for inventing the laser passed him by. This wasn't the only time he fell just short of the credit for a major discovery.

He had independently predicted the existence of the all-pervasive CMBR and encouraged his protégé at Princeton, Peebles, to perform more detailed calculations. Peebles's calculation, in the context of the hot big bang model, predicted that the fireball start of the universe had produced a relic background of radio waves. This imprint of the big bang, he claimed, would fill all space uniformly and have a steady, detectable temperature of roughly ten kelvins, even "as low as 3.5" kelvins.[12] He noted further that the CMBR would appear as a constant hiss in a sensitive radio telescope. Peebles and Dicke were unaware of Gamow, Alpher, and Herman's relevant calculations and predictions that had been published in the 1940s. They were rediscovering the CMBR.

At the same time that Penzias and Wilson were puzzling over their strange hiss and Dicke and Peebles were thinking about the CMBR, two Russian physicists, Igor Novikov and Andrei Doroshkevich, entered the fray. They had read the papers from Gamow's group and revisited their calculations, not only refining them but also making concrete observational forecasts for the detection of this radiation, a relic of the big bang. Despite being theorists, Novikov and Doroshkevich estimated the feasibility of detecting the predicted faint signals. In 1964 they published a brief paper in *Soviet*

Physics Doklady claiming that the microwave background predicted by Alpher, Herman, and Gamow was detectable and that the specifications of the sensitive Crawford radio telescope at Holmdel were ideal for its discovery. Of course, as luck would have it, Penzias and Wilson were, alas, unaware of Novikov and Doroshkevich's paper. Communication between scientists in the United States and in the Soviet Union was not particularly active or efficient at that time. Besides, there were gnawing divides between theorists, observers, and experimenters.

Penzias, meanwhile, contacted Dicke and invited him to come see the apparatus at Holmdel to try to explain the origin of the mysterious and persistent hiss. Dicke brought his entire group along. Two of Dicke's other protégés at Princeton, Peter Roll and David Wilkinson, both experimental physicists, were at that time putting the final touches on their own radio receiver to search for the CMBR as predicted by Peebles. Following Peebles's theoretical calculation and prediction, Dicke's group was engaged in the quest for this primordial signal. The CMBR was the Rosetta stone of the universe, the signal that was the key to deciphering the origin of the cosmos. It's not hard to imagine the mixed emotions that Dicke must have felt while looking at Penzias and Wilson's data: the exhilaration at their momentous discovery and the simultaneous deep disappointment at being scooped when he and his group were so tantalizingly close to detecting it themselves. He was immediately convinced that the signal that Penzias and Wilson had detected was the CMBR and that it was strong evidence for the hot big bang model. Penzias and Wilson, however, were skeptical of this model—and of cosmology in general—because of the raging controversy between it and the steady state model. They decided to present their experimental findings in a paper without any discussion of the theoretical implications. Though they were not sold on the hot big bang model, they were happy to report their findings without explicitly tying themselves to a model that was still tentative in their minds. Besides, the cautious Wilson, to whom the steady state model still appealed, felt that since they had only one data point, it was best to leave any discussion of a model out of the publication of their observational result. Their detection provided one data point, the first one, on the predicted smooth blackbody curve of the CMBR.

So by agreement, the two groups published two separate papers back to back: one by Penzias and Wilson reporting the discovery and the data, and one by Dicke and his collaborators interpreting the CMBR as relic radiation from the big bang. Ironically, it was one of Dicke's patented inventions, the radiometer, a device that uses clever electronic circuitry to detect extremely weak radio signals by separating them from the background, installed on the Crawford radio telescope, that enabled the detection of the CMBR. Dicke had come precariously close, from both the theoretical and the experimental point of view, to making this remarkable discovery. It was yet another tragic near miss of major acclaim for him. In 1978, Penzias and Wilson won the Nobel Prize for their serendipitous discovery of the relic radiation from the moment of creation of the universe. And of course, had Dicke been aware of Alpher and Herman's prior work, he might have started the search for the CMBR sooner and might have well shared the Nobel.

On that fateful Friday, March 26, 1965, when these two groups, which worked hardly thirty kilometers (about nineteen miles) apart, finally met, no one in either was aware of any of the earlier published calculations about the CMBR by McKellar or Alpher and Herman. Counterfactual speculations are always intriguing, and such what-ifs are exciting to explore. The discovery of the CMBR is definitely a case where if matters had been even slightly different, the lives of many individuals would have taken completely different turns. Intellectually, an earlier discovery would have established the hot big bang model sooner and ended the belabored strife with the steady state model.

Personally, the constellation of who received the ultimate accolade—the Nobel Prize—might also have been entirely different.

Despite its innocuous title, Penzias and Wilson's one-page paper "A Measurement of Excess Antenna Temperature at 4080 Mc/s," published in the July 1, 1965, issue of the *Astrophysical Journal*, reports one of the most important breakthroughs of the twentieth century.[13] With the announcement of the discovery of the CMBR and the accompanying paper with the interpretation by Dicke and his collaborators, astronomers' and astrophysicists' view of the hot big bang model transformed practically overnight. Although he interpreted the CMBR as relic radiation, Dicke's emotional allegiance lay elsewhere than the big bang model. It was widely known that he preferred an oscillating universe—a variant of the hot big bang model proposed in the 1930s by the Caltech cosmologist Richard Tolman—because he found it philosophically problematic to deal with the idea of a universe created presumably from nothing in a single instant. Because the heating and cooling in the oscillating universe model are episodic, it is compatible with the production of a CMBR at the start of a cycle of expansion from a hot, dense state.

The excitement around this remarkable discovery was palpable, and the *Astrophysical Journal* leaked news of the CMBR's detection to the press. On May 21, 1965, it appeared in the headline "Signals Imply a 'Big Bang' Universe" on the front page of the *New York Times*. Although Dicke was comfortable with the expansion of the universe, with galaxies flying away from one another in the manner that Hubble had observed, he much preferred them to eventually fall back together again. The science writer Walter Sullivan reported, "Dicke and others would like to see an oscillating universe come out triumphant." As we have discussed before, notably in Einstein's case, the emotional hold of certain scientific ideas can be hard to escape for some individuals. Maybe Dicke's attachment to the oscillating universe held him back from looking earlier and more assiduously for evidence that we now understand was the key confirmation of the hot big bang model. For the cosmological skeptic Wilson, the appearance of this story on the front page of the *New York Times* was apparently proof that the world really was taking cosmology and his and Penzias's discovery of the CMBR seriously. He was right—cosmology was slowly coming of age, graduating from being a speculative and wishy-washy subject to one whose theories could be confirmed and tested with measurements.

The discovery of the CMBR is one of the cases in astrophysics where there was a great deal of unhappiness and discontent over how it was credited. Alpher and Herman, who made the original prediction (despite being slightly off), felt very left out indeed. They had never received any recognition. Once again, I argue that partly to blame was their choice to publish in the *Proceedings of the National Academy of Sciences,* not a journal that astronomers and physicists typically read or routinely submitted papers to for publication. As we saw with Lemaître, publishing a breakthrough result regarding the universe's expansion in a less known journal greatly hinders its dissemination and acceptance. Publishing in journals established and read by the astronomy community at large was now crucial for broader propagation of such new research. Despite the specialization into subfields, the entire community respected and read the publications of an agreed-on cadre of publishing outlets. At any rate, Alpher and Herman were extremely upset that Dicke's group had been unaware of their earlier work. In particular, the journal *Physical Review* rejected Peebles's paper with the new calculations for poor scholarship—for not having cited earlier work. "Jim Peebles knew of our work, unless he was incredibly obtuse," said Alpher, who further noted, "Peebles got from us two reviews of his paper on the background radiation well in advance of the Dicke-Peebles-Roll-Wilkinson 1965 publication." It turns out that Alpher and Herman were the referees who rejected Peebles's submission. As for Dicke, apparently his excuse to Alpher was that his own work had been similarly ignored in other instances and that this was par for the course in the field. Alpher was dissatisfied with this explanation. He believed that Dicke's Princeton-centric attitude essentially meant "if it wasn't invented here, it wasn't invented." By way of recompense, Peebles and Wilkinson published an article in *Scientific American* duly acknowledging the earlier theoretical work by Gamow, Alpher, and Herman, but in the eyes of Penzias and Wilson, it gave insufficient credit to the experimental detection. And that left the Holmdel pair sour.[14] Conflicts over credit for the discovery of the CMBR left deep scars for the key early players—Gamow, Alpher, and Herman. Some attempts at reparations were made decades later. In 2005, President George W. Bush awarded Alpher the National Medal of Science, one of the highest civilian honors in the United States, with the citation "For his unprecedented work in the areas of nucleosynthesis, for the prediction that universe expansion leaves behind background radiation, and for providing the model for the Big Bang theory." He finally got his due. At the award ceremony, on July 27, 2007, his son, Dr. Victor S. Alpher, accepted the medal on his behalf.[15] Alpher died only a few weeks later, at the age of eighty-six. Gamow, meanwhile, had died in 1986, discontented at not having been recognized for his work on the early universe. He is one of the tragic heroes of the CMBR story.

Prior to the discovery of the CMBR, the physics community had accepted neither the steady state theory nor the hot big bang theory. The steady state theory compensates for the expansion of the universe with a claim of continuous matter creation, because of which the universe appears the same for all time, without a beginning or an end. Each theory had its loopholes (and some gaping holes as well), but the discovery of the CMBR completely shifted their assessment. The big bang theory predicted this relic radiation, whereas the steady state theory neither predicted it nor could it account for it once it was found. Within about two years of Penzias and Wilson's discovery, the steady state theory fell out of favor completely, although Hoyle and a few other loyalists continued to try to incorporate the CMBR into improved versions of this model. Ultimately they failed to do so, and the CMBR turned out to be the pivotal piece of empirical evidence that led to the collapse of the steady state theory.

After detection, the next step for the CMBR was measurement—more detailed investigation and filling out the predicted blackbody curve with data. As this cosmic relic radiation is a blackbody, its spectrum, the shape of the curve describing the distribution of total energy over a range of wavelengths, is theoretically well known. To test if the CMBR indeed follows a blackbody curve, independent measurements needed to be made along several wavelengths. The time had come to go beyond mere detection and discovery, to go beyond the first and only data point, to careful measurements. Precise and accurate measurements of the CMBR required new, more sophisticated tools. Once the idea of the CMBR as a relic of the big bang was fully accepted, given its expected blackbody form, a plethora of detailed, testable theoretical calculations could be made. These in turn greatly strengthened the case for the big bang and provided a deeper understanding of the early universe. The energy in the radiation of the CMBR exceeds that from all the starlight in all the galaxies combined and accounts for 99 percent of the total radiation in the universe.

Despite this successful modeling of the early universe, Gamow's ultimate question—how the heavier chemical elements came into existence—remained unanswered. Today we understand that the dense fireball of creation—the "bang" that produced the soup of matter and radiation—started cooling as the expansion began. Radiation predominated over matter at this epoch—the number of photons vastly surpassed that of any other kind of particle. Protons and neutrons first assembled from their constituent quarks, and within the first three minutes, nuclei of helium, deuterium, and lithium began to form from the neutrons and protons, through fusion. Because fusion requires a very high temperature and density—one of the handicaps that curtails our ability to replicate it in the laboratory—once the universe cooled below a certain temperature, this process stopped, choking the formation of any further heavy elements. Nothing heavier than lithium, the seventh element in the periodic table, could form in the primordial universe. In fact, this was the trick that Gamow, Alpher, and Herman missed in their calculations: every naturally occurring element heavier than lithium was produced in nature's fusion reactors—stellar interiors—much later in the universe, and not in the primeval fireball.

Around 380,000 years or so after the big bang, positively charged nuclei began combining with electrons to form electrically neutral atoms. The creation of atoms allowed matter and radiation to split and begin evolving separately in the universe. We call this divorce of matter and radiation "decoupling." Free of the pressure exerted by radiation, matter began to clump because of gravity, forming the stars and galaxies that we see today. Radiation's first great escape is what we see in the CMBR. The CMBR, which fills the entire universe, is therefore a snapshot of the conditions in the universe 380,000 years after the beginning. Observations from the COBE satellite confirm that the universe was a nearly perfect blackbody at decoupling. Its careful measurements of minuscule variations in the CMBR have provided further support for theories of how galaxies, filled with dark matter, formed and evolved in our universe. The temperature of the CMBR—actually, the very small variations in it, tiny differences across the sky of a millionth of a degree—encapsulates information about all the galaxies and other structures that it has permeated on its journey to us.

Once Penzias and Wilson's discovery was published, the race began in earnest to measure the CMBR at other wavelengths, to map out the full blackbody curve. By the late 1970s it was a hot research topic, and experimentalists at several institutions launched balloons to study the radiation and circumvent the effects of earth's atmosphere that essentially blocked out certain frequencies. A group at Berkeley, for instance, focused on making measurements at short wavelengths, of two millimeters or less, that could help discern the shape of the CMBR and determine if it was an exact blackbody or just approximated to one. The balloon experiments that were needed to probe the range of frequencies were tricky to execute. The apparatus was extremely fragile; each balloon was made of plastic less than a thousandth of an inch thick that could tear easily. These had to be filled with helium and then released with the detectors for measurement attached to the bottom. Of course, this was not all. Once up, all the relay electronics had to work, and the measurements needed to be beamed down to earth for analysis. There were also transatlantic rivalries—a British group based at Queen Mary College was working on similar measurements with radiometers attached to balloons in the 1970s. It was a heated chase.

The instrument that catalyzed the discovery of the CMBR was Dicke's radiometer. He had worked out its basic principles by realizing that the noise that heat produces in a resistor in an electrical circuit is a direct measure of the temperature of the resistor. So placing such an apparatus in a sealed cavity with an antenna that captured radiation from the outside would give a researcher a sensitive thermometer. The "antenna temperature" could be measured simply by looking at the temperature of the resistor, with which it aligned and matched.[16] Improved versions of the original 1946 radiometer, spurred by World War II–born interest in radio devices, are what were deployed in the CMBR balloon experiment payloads in the 1960s and 1970s. Slowly, though, it became clear that the detection of the entire blackbody spectrum was necessary, and this was a task best accomplished from space, above the messy confusion that earth's atmosphere causes. The push for a satellite to map the CMBR began with the announcement of an opportunity from NASA that Mather and others responded to with a proposal for COBE in 1974. The experience with X-ray observations from space with the successful launch, deployment, and data collection of the Uhuru satellite, whose mission lasted from December 1970 to March 1973, inspired the opening of newer windows onto space—in this case, the microwave window.

The detection of the CMBR by Penzias and Wilson and subsequent confirmation by many other groups with balloon experiments and other ground-based detectors provided powerful confirmation of the big bang theory. In the past twenty-six years, since the COBE satellite first made possible a precise measurement, WMAP (the Wilkinson Microwave Anisotropy Probe, named after Dicke's protégé, who died unexpectedly in 2002) and the European Planck satellite (named in honor of the German physicist Max Planck), whose mission recently concluded, have confirmed the larger picture of the universe that has left an indelible imprint on the CMBR.

These new results have gone well beyond just validating the hot big bang model, to testing models of elementary particles and the overall paradigm for the formation of structure in the universe. The current accepted theory for the formation of all structure in the universe is predicated on the growth of small fluctuations in the matter distribution at very early times. In our dark matter–dominated universe, gravity amplified these tiny initial fluctuations or disturbances, which generated mass clumps that eventually hosted the formation of the first stars and the first galaxies. The clumpiness of the matter and the subsequent coalescence that formed galaxies left an imprint in the CMBR. Variations of temperature at the minuscule level of one part in a million correlate with all the matter that the CMBR has encountered en route to us today on its long cosmic journey. The missions that have followed COBE have detected and measured these incredibly tiny patterns of hot spots and cold spots with high precision. In addition to this splotchy pattern, the CMBR has several other signatures that prove its primordial origin. We are now talking about the smallest of variations. The CMBR is extremely uniform, so any two opposite points in the sky will differ in temperature only in the fifth decimal place and beyond. The physicist Dennis Sciama predicted that because of the motion of the earth through the CMBR, the Doppler shift (as we discussed for sound and light when describing Hubble's work) will make the radiation's temperature appear slightly higher—on the order of about one part in a thousand—in the direction of our motion. Conversely, the temperature will be slightly cooler in the opposite direction, in our wake. Therefore, there are predictable departures from uniformity in the temperature distribution—which are measurable. In some sense this is reminiscent of the concept of ether, through which the earth and the solar system were supposed to be in motion. The CMBR could be thought of as the cosmic "ether" filling the entire universe. The specific temperature variation across the sky due to the earth's motion is referred to as the dipole anisotropy. Even in the late 1970s it was extremely challenging to detect evidence for the dipole, as the cutting-edge technology was earth-based telescopes and balloons. This motion of the earth needed to be securely measured, because only then could the other sources of nonuniformity—which held the key to ratifying the entire cosmological edifice—come into view.

Enter Mather's COBE satellite, the response to a call from NASA for scientists to propose space-borne experiments. The development and use of technically advanced instrumentation went hand in hand with the complexity of the theory that needed to be validated. The detailed theories of the formation of structure culminating in the galaxies that we now see in the universe were developed at this time. The physicists Yaakov Zeldovich and Robert Harrison independently predicted that the evolution of small matter fluctuations during the cold dark matter–dominated structure formation process would leave a trace on the CMBR photons that encountered them while traversing the cosmos. Assembling galaxies would produce small fluctuations in the temperature of the CMBR as it snuck past them. These theoretical calculations could now be extended to greater precision within the framework that Peebles had developed. Therefore once this cold dark matter paradigm got firmed up, more sophisticated and testable predictions could be made. It became possible to characterize the full complexity of the evolution of matter—both dark matter and ordinary matter—and its interactions with the traveling photons with numerical simulations that could then be used to calculate the production of minuscule variations in the temperature of the CMBR. Once again, simulations were an indispensible and powerful tool for discriminating among and validating models in conjunction with observational measurements. As with dark matter, the growing sophistication and raw computational power that were part of the computer revolution proved critical to making the high-precision predictions from the model that needed to be folded into interpretation of the satellite data.

The sensitive instruments aboard the COBE satellite first measured these tiny variations in the temperature of the CMBR, the impact of matter encountered in its journey across the cosmos. I still vividly remember sitting in a cramped and tightly packed lecture theater as an MIT undergraduate when George Smoot, the principal investigator of the Differential Microwave Radiometer on COBE, presented the results in a seminar in 1989. They were awe inspiring. The data points from COBE traced the smooth curve of the blackbody so perfectly that the measurement errors on it were tinier than the thickness of the printer's rendition of the graph that overlay its shape. Physicists were euphoric, and every major newspaper in the world reported the discovery with triumphant headlines, such as the one in the *New York Times,* which read, "Scientists Report Profound

Insight on How Time Began," or that of the *Independent* (London), which ran, "How the Universe Began."[17]

With the detection and subsequent measurements of the CMBR, cosmology, which had been dogged by large uncertainties in the determination of key properties of the universe, such as its age and the Hubble constant, started to become a precision science. Until the COBE mission and the more recent follow-up measurements from WMAP and the Planck satellite, cosmology had the reputation of being speculative and even somewhat crude compared to other, quantitatively sophisticated branches of physics, such as particle physics, where highly accurate measurements down to the fourteenth decimal place are de rigueur. COBE marked a new era of precision cosmology with highly accurate measurements, and it finally became a respectable science. By this time, in the 1980s and 1990s, there was much more synergy between theory and observations as well.

Scientists cannot, of course, manipulate cosmic phenomena as if they were laboratory specimens. Despite this fundamental limitation, this inability to perform controlled experiments, cosmology gained in legitimacy as a quantitative science with the measurements of the CMBR. They caused an explosion in knowledge of the early universe with unprecedented accuracy and have been increasingly refined with every subsequent space mission. The instrumental resolution of WMAP, which was launched in 2001, was thirty times greater than that of COBE, and the Planck satellite (launched in 2009) had a resolution two and a half times higher than that of WMAP and greater sensitivity in a larger number of frequency channels. With the incoming Planck data, our understanding of the subtle ways that matter and radiation interact over the course of cosmic history will continue to grow phenomenally. The current frontier in CMBR measurements is its polarization pattern. Polarization is a property of waves, like light, that can oscillate in more than one direction.

You may have encountered the trick of making a light bulb disappear with a polarizing filter. Holding the polarizer up to the light, one clearly sees a brightly glowing bulb, but rotate the filter by 90 degrees, and the light and consequently the bulb will no longer be visible. Light polarized along one direction gets completely blocked. Polarization is seen even in X rays and microwaves. Incredibly, the polarization pattern of the CMBR can be measured, and this is the current observational challenge at the research frontier. Measuring this imprinted pattern can tell us even more about the infant universe.

As we have discussed before, new ideas and questions demand new tools and instruments, which then get honed further and further. The need for certain skills and expertise to perform these ever more sophisticated measurements of the CMBR has impacted the professional structure and the everyday practice of cosmology. As the historian of science Peter Galison points out, the division of labor in particle physics created new professional "trading zones," which redefined expertise. In the case of cosmology, the study of the CMBR created a trading zone between theory, instruments, and experiments. It took the collaboration of engineers, who built the instruments; scientists, who collected the measurements; and theorists, who interpreted the data—three distinct communities of people—to bring the CMBR missions to fruition. This all started with Dicke and others working at the MIT Radiation Laboratory, itself an extraordinary trading zone, both in terms of knowledge creation and as a physical space where interactions among sectors of the community that were previously sealed off from one another took place. As I have mentioned before, the 1980s and 1990s marked a distinct intellectual stage for cosmology, with instruments driving the division of labor in research. Galison notes that the existence of trading zones transforms not only the practice of science but also how new scientific claims are made and the nature of disputes that arise.[18]

The level of sophistication wrought by technology means that arguments and resistance are no longer reducible to a clash of ideas. This is true for cosmology as well and explains why the nature of resistance within the scientific community that we have encountered in this chapter is so fundamentally distinct from that in the case of black holes, which involved a standoff between Chandra and Eddington, or of the idea of the expanding universe, which saw Einstein finally concede to Hubble's observations.

By the 1990s it would have been hard for a patent clerk, outside the academic research enterprise, to be recognized as a scientific genius. Scientists are now part of a professional cadre, and the training track to become one was standardized over the course of the twentieth century. The individual sense of wonder and curiosity has now been harnessed into parceled-out piecemeal work among members of a large team. But this change in the structure of how scientific knowledge is produced is by no means the end of controversies or disagreements—instead, they are simply articulated and resolved differently than before. The fights now are about the details of data processing, calibration, and whether researchers have appropriately taken into account the various sources of error that contaminate the measurements. Because all the work is implemented in teams, the clashes between collaborations take a more ritualized form of rivalry, although fame, credit, and accolades are still what are at stake.

Consider for example the most recent dispute that emerged in the claimed measurement of the polarization signal in the CMBR. WMAP saw the first hints of polarization. Now we needed a more accurate determination of the strength and significance of the signal, which took into account all the sources of error that could scupper it. In March 2014, the BICEP (Background Imaging of Cosmic Extragalactic Polarization) team, which was making precision measurements of the CMBR polarization with a radio telescope at the South Pole, announced their dramatic discovery. They convened a press conference to report their measurement of the signal with an extremely high degree of confidence. But the swirly polarization pattern seen in the CMBR could easily be contaminated by dust that it has encountered in our own galaxy. There was strong disagreement from the rest of the scientific community about how the BICEP team had corrected for the confounding effect of dust in our galaxy, which is the final sieve through which the CMBR passes before it is measured. The team's characterization of the properties of this galactic dust and its impact on the measured polarization was called into question, most notably by theorists at Princeton. Building on Dicke's legacy, Princeton is still the stronghold of CMBR science. Understanding the effect of dust is critical to establishing the provenance of the polarization signal—whether it has a primordial origin or is just the imprint of dust in our galaxy. In today's globally interconnected world, this dispute and questioning occurred in full public view. This was a new venue for cosmology. All the questioning that used to happen, as we have seen, in the safety of specially convened meetings or conferences, behind closed doors, now played out on the Internet and in social media in an unregulated and unstructured manner. Matters that were once adjudicated among experts, deliberations after which a consensus view was arrived at and presented, the entire unfolding that we have seen in the previous chapters, now happened in a fishbowl. The messiness of the scientific process was laid bare. Opinion within the community of cosmologists was divided over whether this new development was helpful or harmful to intellectual rigor. Of course, in the case of BICEP, more and better data from the Planck satellite settled the matter. The BICEP experiment did overestimate its calculation of the credibility of the polarization measurement. The devil was in the dust, so to speak—in this case in the details of systematic errors in the estimate of the role of dust that led to the spurious signal.[19]

I believe that this open discussion and airing of criticism is a good thing. It opens the window on how science works, demolishing the inaccurate image of research as a clean, objective way to generate fixed truths about nature. It shows that scientific theories are provisional and that scientists as a community carefully scrutinize any new claim and its replicability. It also shows how the process by which new and radical scientific ideas are assessed, evaluated, and accepted has fundamentally changed today.

7

THE NEW REALITY AND THE QUEST FOR OTHER WORLDS

When Edwin Hubble looked through the telescope on Mount Wilson to gather data on cold cloudless nights, little did he know that his insights would transform our knowledge of the cosmos forever. And when Arno Penzias and Robert Wilson were trying to tune out the persistent hiss on their radio telescope, they were completely unaware that they had detected relic radiation emanating from the big bang.

The discoveries I have considered in this book have revolutionized our conception of the universe and our place in it. They have also catalyzed fundamental shifts in the practice of science as an intellectual activity. Following the progression of radical ideas in cosmology from their inception to their acceptance, we have seen how debates about discoveries shifted away from being between individuals. Slightly different obstacles line the gauntlet of scientific acceptance today. The competition now is largely between large teams pursuing scientific questions at the frontier. Although evidence and data propel scientific acknowledgment, serendipity and intellectually influential individuals continue to play a role. But no longer can a single person drive the acceptance of or resistance to a new idea. A single person might have influence, because science is still somewhat hierarchically organized, but I'd argue that intellectual power and sway are more evenly spread among institutions around the globe today than they were even twenty years ago.

Subjectivity and emotional investment remain important, both in how scientists conceive creative ideas and in how others contest or accept these ideas. Science has always been an activity suffused with personal passion and wonder, and this has not changed. Contrary to the image of science as a methodology to extract fixed truths from nature, I have tried to show that it is dynamic and provides only a current map forward—and that map is inherently provisional. Without it we would lack direction, but as it points to terra incognita, it tells us mostly about where we've started from and what remains unknown and as yet uncharted. The twentieth-century tussles in cosmology illustrate the deeply psychological aspects of science—the drive to explore nature as well as the limits that our minds inevitably impose on our understanding.

Over the past hundred years in cosmology, the kinds of questions that scientists tackle and the tenor of intellectual engagement have fundamentally transformed, with individual triumphs of brilliant insight giving way to a more organized, collective effort involving technical and narrowly defined areas of expertise. For instance, the skills and know-how needed to design and launch the Cosmic Background Explorer (COBE) satellite in 1989 to measure the cosmic microwave background radiation (CMBR) are very different from what theorists needed when they made the first prediction of the CMBR's existence forty years earlier. The open questions now are incredibly complex, and we are submerged by data and require the efforts of diverse teams to comprehend them. By the time the automated large-scale supernovae searches were under way in the early 1990s, the amount of data pouring in that needed to be analyzed had increased exponentially. Not only the volume but also the speed of data accumulation is a current challenge. Meanwhile, thanks to advances in hardware and software, our computational capability to model and interpret this deluge of data has grown tremendously. As a result, astronomy has been at the leading edge of the developing big data revolution. As we have seen, the new methodology of numerical simulations has bridged the sharp divide between theoretical and observational work. Research teams are now composed of theorists, observers, simulators, and engineers. The insularity of subspecialties is to a large measure a thing of the past. The confluence of ideas and instruments is stronger than ever.

But unlike a map, the human mind and its capacity for thought and imagination know no boundaries. Science continually fans human curiosity and is driven by it as well. As a pursuit, research encourages constant questioning of the status quo. There are rewards for rebels, the outsiders who mount challenges and can make a persuasive case for a change in world view with data that supports the shift. Good scientists do not take anything for granted. If a scientist finds a problem with a currently accepted idea, model, or conception and can show that there is evidence to support a new claim, then the rest of the scientific community will eventually take notice and reevaluate their stance, as we have seen thus far in this book. This is one of the hallmarks of research—the progressive expansion of questioning and understanding. Because curiosity and wonder power the scientific enterprise, it is no surprise that our quest to comprehend our place in the cosmos is far from over. There are two somewhat related, compelling questions that astronomy and cosmology are in the midst of tackling today. Both have an existential flavor, and they pertain to our uniqueness: whether as a species we are special, and whether our universe as a whole is simply a statistical fluke. At the heart of both these questions lies the deep desire to discern our place in the cosmic context, our place and imprint on this vast map.

In the realms of theory and observation, cosmology has been tackling and successfully answering some of the big questions that Chandrasekhar, Eddington, Einstein, and Hubble faced in the first half of the twentieth century. Our knowledge of the universe—its contents and its fate—compels us now more than ever to examine if there is any way that humans are set apart. Are the sentient beings on our pale blue, rocky planet mere statistical anomalies in the cosmic saga? As we sit with this notion of a rapidly accelerating, unfixed universe, in which galaxies drift farther and farther away from one another, we are daring to speculate and contemplate the existence of other habitable worlds with other beings and, beyond that, even other entire universes. The unbounded universe has unleashed a new boldness. The realm of science fiction was typically where unconstrained speculation and fantasy were unleashed. But the stuff of science fiction is becoming concrete and real. The frontier now is the search for other worlds—the quest for habitable exoplanets nearby and other possible universes far, far away.

Our accumulated knowledge of the universe—our comprehension of where we are today and our understanding of how we arrived here—has informed the current questions. They reflect our continual search for our place in the universe, our role in the grand scheme, and signal our discomfort with the current state of knowledge. Cosmic discoveries to date have left us unmoored, and the dizzying pace of scientific change has disoriented us. We have found our significance continuously diminished, a species inhabiting one planet among eight (previously nine) in one solar system among several thousand others, in a galaxy among several billion others growing increasingly distant. Our primal drive to situate ourselves underlies the quest for habitable planets in our vicinity and speculation about the existence of other universes. While the deeper impulses driving these two questions are one and the same, they are scientifically quite distinct and require vastly different methodologies and lines of attack.

In the quest for nearby habitable planets, although scientists have hoped to find variegated exoplanets in our backyard, we have particularly searched for those that are most similar to earth. We hope that these are the most viable places to host life, specifically intelligent life of the familiar kind. The quest for such habitable planets is under way. NASA's Kepler satellite has returned an unexpectedly rich haul of candidates around nearby stars.[1]

The other radical idea that we are now examining lies in the realm of theory and mathematics: the multiverse. This is the notion that our universe might be just one of an ensemble of many. We have seen in this book how bizarre and wonderful this universe of ours is, one whose secrets we have slowly unraveled in the past hundred years. Einstein's field equations helped formulate the connections among the geometry, contents, and fate of our universe, and astronomical observations over the past century have established its particular trajectory. We have now homed in on the model solution that best matches cosmological data, one with relentless and accelerating expansion.

As the eminent cosmologist Martin Rees elaborates, it turns out that to fully specify all the relevant properties of our universe, surprisingly all we need to know are just six numbers! We have determined them all empirically. These pivotal numbers, called the cosmological parameters, are *N*, which has a value of 10^{36} and measures the relative strength of electrical forces to that of the force of gravity between atoms; epsilon, which has a value of 0.007 and defines the strength of binding of atomic nuclei; omega (which we encountered in chapters 4 and 5), which has a value of 1 and is a measure of the mass-energy content of the universe; lambda (which we encountered in chapters 2 and 5), which is the cosmological constant and has a value of 0.7; *Q*, which has a value of 10^{-6} and measures the strength of the initial fluctuations that seeded the formation of all the stars and galaxies in the universe; and finally *D*, the number of spatial dimensions in our universe, which is three.[2]

If the values of these cosmological parameters deviated slightly, even by a few tenths of a percent, from their measured values, then we would not exist! No humans. No earth. No universe. Life on earth would not have been possible, and all of what we know about

The Cosmological Parameters

PARAMETER	DEFINITION	MEASURED VALUE
N	Ratio of electric force to gravitational force between atoms	10^{36}
ε (epsilon)	Binding strength of atomic nuclei	0.007
Ω (omega)	Total mass-energy content of the universe	1.0
Λ (lambda)	Cosmological constant	0.73
Q	Size of initial fluctuations	10^{-6}
D	Number of spatial dimensions	3

the cosmos would have been unknowable. For instance, if *N* had been just slightly smaller, then the universe would have been so short lived and tiny that there would have been no time available for biological evolution beyond the most rudimentary creatures on earth. If epsilon had been off by .001, then the synthesis in stars of elements in the periodic table beyond lithium would have been impossible. No organic life that we know of would be here. We have already seen what fates different values of omega and lambda imply—a crunch or expansion of the universe, ending our story before it began. The variable *Q* is also perfectly poised: if it were any larger, then the universe would be too violent for stars to survive, and if it were any smaller, then all the structure that we see could not have assembled. If the dimensionality, *D,* were 2 or 4, life as we know it simply would not exist.[3]

All of this of course carries a teleological whiff. There's even a name for it: the anthropic principle, which is the philosophical claim that observations of the physical universe ought to be aligned with the fact that conscious life exists. That is, these precise values for the cosmological parameters have permitted us to be here and ask the question in the first place. What is clear is that the universe has not singled out carbon-based life to exist; in fact, a more compelling argument states that the apparent fine-tuning of our universe is likely a result of selection bias, because only a universe that can support life might give rise to beings who can ponder and question the matter. Can we figure out why these numbers have the values that they do? In any attempt to more profoundly understand, say, why omega has its specific value for our universe, when it could have had just about any other value—say 0.001, 0.1, 10, or even 42—the logical conclusion is that we simply happen to inhabit a universe where this is so.[4]

If it is merely selection bias, we have only one universe in which we can make measurements and by definition cannot conclude why this is so while sitting inside it. But should we instead be open to the possibility, by extension, of universes where these six crucial parameters have completely different values? In that case the cosmological parameters would have the values that they do only in our universe, and our universe would constitute only one realization of many potential universes with their own combinations of values for these parameters—referred to as bubble universes—that could in principle exist and together make up what is called the multiverse. Inevitably, this implies that there could be an infinite number of bubble universes out there, each with its own hexad of values for the cosmological parameters.

If we adopt the probabilistic view that other possibilities with associated degrees of plausibility could exist, then the straightforward conclusion is that we have a particular combination of six numbers and that the actualization of an infinite number of other possible combinations is likely. Of course, other values of those six numbers would give rise to vastly different universes, with diverse geometries, exotic contents, and alternate fates. The probabilistic view can free us from the need to invoke the anthropic principle and to account for the fine-tuning required to explain why these particular values for the six numbers, and why us. We can therefore cleanly circumvent the question of why our universe has these specific values for the cosmological parameters by using the argument that it is just one of many realizations out of all possibilities—in fact a single constituent of the infinite multiverse. Each of these possibilities could be realized, and there could be an infinite number of so-called bubble universes floating around with other combinations for the cosmological parameters, each beginning with its own big bang.

How might we test the hypothesis of the existence of bubble universes? Well, the finite speed of light allows us access to only a portion of our universe, let alone other universes that may be well beyond. A billion years from now, the visible horizon of our universe will be larger and a bigger portion of our universe will have come into view, because light will have reached us from objects that are a billion light-years farther away than the visible edge today. If light cannot open up the view beyond our universe even in a billion years, then how can we possibly conceive of making measurements and observations of other universes?

In this speculative realm on the largest imaginable scales, concepts such as the multiverse have brought to the fore an intellectual challenge of a new kind: the necessity to accept an explanation supported not by directly testable theories but by extrapolated versions of currently accepted theories. We seem to have hit the current limit of scientific explanation, which it appears the foreseeable invention of more advanced instruments will not remove. Perhaps what we need are theoretical reconceptualizations.

This might, though, be a fundamental limit because, on the one hand, it involves physical regimes where the validity of theories might be simply impossible to test. On the other hand, to take a lesson from the history of cosmology, could Nicolaus Copernicus, when he wrote *De revolutionibus orbium coelestium* in 1543, have foreseen our ability to land on the moon in 1969 and bring back samples to the earth to study? Or the landing in 2014 of the probe Philae on the surface of the comet 67P/Churyumov-Gerasimenko? Probably not. Neither could he have foreseen the invention of spectrographs and cameras that have provided exquisite images of the distant universe that he did not even imagine could exist. Therefore, at the moment we may not know whether the multiverse concept is testable, but there is no reason to believe that this will remain the case a couple of hundred years from now, if not sooner. It would be arrogant of us to attempt to predict the course of future science. What we can and ought to do instead is give free rein to our imaginations and see what rich and new possibilities may open up.

How can we even begin to tackle this daunting question in familiar explanatory terms within the frameworks of physics and mathematics that we have developed? Part of the challenge is connecting a description of the pre–big bang universe to the universe that we can see and measure. String theory is the branch of physics that endeavors to do just this. It conceives of particles in the universe as generated by oscillations of strings, like those in musical instruments, during the pre–big bang era. Imagine a violin string tuned by stretching it under tension. Varying the tension produces different musical notes, which can be thought of as the excitation modes of the string. Similarly, in string theory all elementary particles produced in the universe that we detect today can be conceptualized as musical notes produced by the elementary strings that existed prior to the big bang. And of course the elementary strings will have needed to be plucked in order to be excited. The analogy with violin strings goes only so far, however, because theoretical strings are not anchored to any instrument! String theory offers a mathematical prescription that renders pre–big bang calculations tractable. Scientists currently working on the problem of particle creation at the interface of cosmology and string theory are looking fervently for potential observational signatures of the multiverse—perhaps some rogue ripples in the CMBR that might have resulted from the collision of our universe with another. We hope for a radical shift in understanding that might reveal some detectable, measurable evidence of the multiverse in our universe.

Returning to the six defining cosmological parameters, we need to keep in mind that although these numbers could have different values in other parts of the multiverse, the presumption and working hypothesis is that the physical laws as we know them are unaltered and valid everywhere, even in other universes. There is no credible reason to presume that there might be universes that dance to the tune of entirely different physical principles and where, for instance, the forces that are familiar to us do not exist.

Aside from preventing us from having to resort to anthropic arguments, recent progress in understanding the very early quantum universe suggests that the processes that might have evolved the initial conditions for our universe are rather general and could therefore easily spawn other independent bubble universes. There is a particular version of string theory, the landscape idea, in which many bubble universes can be generated naturally. As a theoretical idea and speculation, it is rather appealing, but again, it raises some questions, such as whether the multiverse is a testable scientific idea. Might string theory provide the ultimate breakthrough needed to comprehend how other bubble universes could conceivably be generated? We will just have to wait and see.[5]

Looking back at the progress that we have made in cosmology even in just the past hundred years suggests that we have every reason to be optimistic. The current thinking is that there were many big bangs because every bubble universe began its existence as and emerged from its own primeval fireball. That is, the process that originated our universe might operate in the same way for the rest of the bubble universes that constitute the multiverse yet provide different outcomes for their geometry, contents, and fate. What empirical proof of this we might be in a position to gather is completely up in the air right now, although cosmologists are busy trying to figure out if there could be any detectable signatures of the interaction between and collision of two bubble universes. Attempts are currently under way to simulate what the emergence, evolution, and collision of bubble universes might produce. We are poised at the precipice—there may be a conceptual breakthrough, we might never really know if the multiverse idea is correct, or we may figure out that it is impossible to ever know.

Even if we were open to the probabilistic view and therefore to many alternate universes, we are in the dark about the allowed ranges for each of these six numbers, which would circumscribe the entire suite of possibilities. The most intriguing option, though, in my opinion, is that other universes might permit new and different physical laws apart from those that are familiar to us. The concept of the multiverse offers an unprecedented epistemological challenge: perhaps the need to embrace a radically new kind of explanation and proof that may never involve direct measurements but may rest on consistency tests of extrapolated versions of securely established theories. The question of how unique our universe might be is not one that we are likely to settle soon.

The odds, however, for settling the other big question, debated since antiquity, are higher. I argue that we are likely to know relatively soon if we are alone in our universe. Let us take a long view and trace the history of when and how we became open to the idea of other worlds beyond our own in our universe. This possibility, like many others we have encountered in this book, occurred first as a leap of imagination before crystallizing into a question that could be investigated scientifically. Imagination often provides the impetus for and brings to fruition ideas that are worth exploring and finding proof for. The idea that multiple worlds akin to ours might exist dates at least to the ancient Greeks, those inveterate explorers of nature who laid the groundwork for much of our current knowledge map. The oldest piece of surviving evidence of this speculation dates to the fourth century BCE, although records of earlier discussions do exist. Epicurus, referred to as an atomist, had a materialist—atomist—world view and wrote boldly in a letter to the famed historian Herodotus about the possibility of other worlds: "There are infinite worlds both like and unlike this world of ours. For the atoms being infinite in number, as was already proved, are borne on far out into space. For those atoms, which are of such nature that a world could be created by them or made by them, have not been used up either on one world or a limited number of worlds. . . . So that there nowhere exists an obstacle to the infinite number of worlds."[6]

This claim harks back to an even older deliberation that originated in ancient Greece prior to the atomists. Cosmic pluralism is the philosophical belief in the existence of many worlds (potentially an infinite number) in addition to earth that may harbor life. The debate around it truly began in 600 BCE with the pre-Socratic father of all Western philosophy, Thales of Miletus.[7] The entirely theoretical argument then was not about the existence of other worlds—what we think of as solar systems today—or specifically life-bearing planets but rather centered on grappling with the notions of infinity and the edge of the cosmos. Thales and his student Anaximander—whom we encountered in chapter 1—as early proponents of science, in fact proposed an infinite cosmos. The atomists Democritus and Epicurus ascribed to this view. Coming after them, Plato and Aristotle stood in strong opposition to this idea and claimed instead that the earth is unique. The Aristotelian picture meshed rather well with later Christian belief, and this led to the suppression of the idea of pluralism for almost an entire millennium. Stories revolving around cosmic pluralism abounded in the Middle Ages, though, and were rife in the cultural imagination of the period. For instance, several Islamic scholars in what is now Iran richly imagined and speculated on these other worlds.

The Islamic scholar and intellectual Fakr al-Din al-Razi, who lived from 1149 to 1209 and wrote extensively on medicine, astronomy, and physics in the context of the Qur'an, described his model of the physical world in his book *Matalib al-aliya* (The sublime requests). In it he criticizes the geocentric view and explores the notion of many distinct worlds besides ours. He questions the interpretation of the term *worlds* in a Qur'anic verse and speculates on whether it refers to multiple worlds inside the cosmos or to actually many other worlds beyond. Al-Razi challenged the Aristotelian notion of a single cosmos and a single world around which all revolved. His rejection of geocentrism arose from his belief in atomism, according to which atoms move, combine, and separate continually in empty space. Knowledge exchange and transfer between the Greek and Islamic worlds was ongoing at this time, so it is not surprising that al-Razi was aware of atomism. He also discussed the notion of voids—empty spaces between stars and constellations in the universe—that contain very few or practically no stars and argued that there exists an infinite outer space beyond the known world, which God has the power to fill with an infinite number of worlds.[8] In some sense, our current willingness to speculate in a completely unbounded fashion on the possibility of entire universes might well have its roots in al-Razi. In the Middle Ages, the Arab world depicted cosmic pluralism in fiction as well, sumptuously. In the compendium of fables *One Thousand and One Nights* (also known as *The Arabian Nights*), the tale "The Adventures of Bulukiya" describes many fantasy worlds, each filled with unique sets of life-forms. In this context, life everywhere was a sign of God's omnipotence.

Although today we are seriously examining these speculations on life elsewhere, this was not the case in the 1500s, and those who dared to dream and speculate in this vein were sometimes labeled heretics and even put to death. One such was the Italian mystic and philosopher Giordano Bruno. For many grave transgressions against the church, including his so-called flights of fancy about the existence of other worlds, he paid with his life. Born in 1548 in the town of Nola in Naples, to the soldier Giovanni Bruno and originally named Filippo, he swapped his given name to Giordano on entering the Dominican order. Attracted as a young man to powerful new ideas, Bruno challenged the order and had to flee persecution—in fact, he ended up fleeing from one place to another his entire life—moving from France to England to Germany and finally to Venice. He harbored many beliefs that were deemed heretical, among them his theories of the infinite universe and the multiplicity of worlds. He also rejected the geocentric view and leaped even further beyond it, propelled not by mathematics but entirely by his intuition. He dispelled the classical notion of a finite universe enclosed within a sphere of fixed stars, challenging some of the assumptions of the new and then radical Copernican view. For Copernicus, although he replaced a geocentric with a heliocentric view, the cosmos consisted of the solar system and the fixed stars alone—it was bounded, and Bruno saw no such boundaries. Of course, it is important to note that Bruno would have been burned at the stake even if he had not espoused the idea of an infinite universe or the existence of many other worlds, as it was his denial of the divinity of Christ and the virginity of Mary, plus other pointedly heretical beliefs, that put him in the church's cross hairs.

In some ways, Bruno's vivid imagination and speculations anticipated modern science. The well-hewn cobblestones in one of Rome's famous squares, the Campo de' Fiori, bore silent witness more than four hundred years ago to his gruesome and macabre silencing for propounding radical ideas that were contrary to accepted religious beliefs. On the unrelentingly cold day of February 17, 1600, Bruno was burned at the stake in the square with his "tongue imprisoned on account of his wicked words."[9]

The Catholic Church had tried him and declared him a heretic, and he was served a death sentence. Bruno had published his unorthodox views in 1584 in *De l'infinito universo et mondi* (*On the Infinite Universe and Worlds*), which the church placed in 1603 on its long list of books banned for heresy or blasphemy, the *Index librorum prohibitorum*. Since the church at that time tightly regulated the limited number of printing presses, books that ended up on this list typically were not openly in circulation and therefore not as widely read.

Despite this backdrop, there soon came a dramatic shift that resurrected these wild possibilities and caused further clashes between the church and proponents of new ideas. It was sparked by the invention of the telescope—a transformative instrument. Repurposing a spyglass, Galileo Galilei pointed it skyward in 1609. His telescope suddenly magnified and brought distant vistas into view, opening up many new questions about what lay beyond the earth. The revelation of new celestial objects suddenly made the consideration of the possibility of other worlds admissible again. This also marked the beginning of the end of naked-eye astronomy. It was, however, only during the late Enlightenment that many philosophers and writers in Europe made cosmic pluralism a mainstream discussion.

In 1686 the French polymath Bernard Le Bovier de Fontenelle published *Entretiens sur la pluralité des mondes* (*Conversations on the Plurality of Worlds*). This was one of the most loved classics of the early French Enlightenment. The book is a set of imaginary conversations between a philosopher and his hostess, the Marchioness, that take place during their walks through moonlit gardens while they survey the glittering night sky. Through these informal conversations, loaded with his rich imagination, Fontenelle describes the new order of the cosmos—the Copernican world view—with clarity and simplicity. For what is surely a remarkable decision for its time, he makes the interlocutor a woman, inviting female participation in the otherwise exclusive male domain of scientific discourse. The pedagogical concept of the fifth evening, which appears as that chapter's subtitle, is "Every fixed star is a sun, which diffuses light to its surrounding worlds."[10] In this chapter, the philosopher explains at length the possibility of other worlds—planets that revolve around other stars—and even life-forms on them.

Continuing this French tradition two centuries later, deeply inspired by Fontenelle, the astronomer Camille Flammarion, who strongly believed in the existence of other worlds, began to write about the idea. He authored more than seventy books and was a great science popularizer in his day. Starting out as an astronomer at the Paris Observatory, he eventually set up his own observational facilities in the suburbs by 1883. His research work involved detailed telescopic observations of the lunar and Martian surfaces and the properties of stars. His first book, *La pluralité des mondes habités* (*The Plurality of Inhabited Worlds*), published in 1862, boldly expounds his view on the possibility of life elsewhere and established his reputation in the public sphere as a leading advocate of cosmic pluralism. His ideas were deeply influential, and by 1882 many of his books had been reprinted up to thirty-three times and translated into multiple languages. With J. H. Rosny, Flammarion proposed the existence of truly alien beings that were markedly different from humans, not just minor variants. In his *Les mondes imaginaires et les mondes réels* (Real and imaginary worlds; 1864) and *Lumen* (1887) he speculates even further, describing exotic imaginary plant species, that are not only perceptive but also breathe and digest. Flammarion's belief in extraterrestrial life forms arose from his idea of the universal existence of transportable (transmigratory) souls that can inhabit flora and fauna alike. Yet again, his imaginative leaps, like Bruno's, were not powered by any scientific calculations. He considered humans to be "citizen[s] of the sky" and other worlds "studios of human work, schools where the expanding soul progressively learns and develops, assimilating gradually the knowledge to which its aspirations tend, approaching thus evermore the end of its destiny."[11]

Flammarion's best seller, though, was the widely translated *Astronomie populaire* (*Popular Astronomy;* 1880), in which he passionately argues for the existence of life on the moon and on Mars. Bolstered by the Italian astronomer Giovanni Schiaparelli's claim of canals on the surface of Mars, he made the case for not just life but intelligent life and an advanced civilization there. He even surmised that the Martians may well be a race superior to our own.

The lively debate about life on Mars, including its potential and possible signatures, has continued vociferously to this date. It has been tantalizing to consider, as in many ways Mars is remarkably similar to the Earth and is the closest planet to us. The length of its day is similar to ours, as are the durations of its seasons. The Martian surface is pristine and has remained unchanged, since the planet has no tectonic activity. Searches for evidence of life there began in the nineteenth century, and they continue today through telescopic investigations, flyby probes, and now landing missions, including the most recent NASA rover, Curiosity, which touched down on the red planet on August 6, 2012. Curiosity has found evidence of ancient water on Mars, water that has long since evaporated. Drilling through old Martian rocks, it also found traces of the organic molecule methane, but no life and certainly no intelligent life. It remains an open question whether any life-forms currently exist on Mars or have existed there in the past.

So where do we stand on the issue today? A 2013 poll conducted by the *Huffington Post* and YouGov.com found that 50 percent of Americans think that there is some form of life on other planets, about 17 percent think there is not, and the remaining 33 percent are unsure. Respondents were more skeptical when asked if they believed that intelligent life had visited our planet. Among those who indicated a belief that life exists on other planets, 45 percent asserted that aliens have visited earth.[12]

Regardless of the vagaries of public opinion, scientific searches are currently under way to find habitable planets around other stars—exoplanets. There has been a phenomenal yield, and part of this recent success is owed to the instruments aboard the Kepler space telescope.

We are luckily alive at a time when the answer to the age-old question of whether we are alone in the universe appears to be within reach. In the past fifty years we have landed on the moon, left human footprints on the lunar surface, sent several probes to other planets in the solar system, had a spacecraft—Voyager 1—exit the solar system, landed the rover Curiosity on Mars for geological exploration, and watched images of the surface of Pluto and Charon taken by cameras aboard the New Horizons spacecraft. We now have remarkably sharp images of several planets and dwarf planets—showing such features as the rings of Saturn, the red spot on Jupiter, the storms on Io, and the heart-shaped region on Pluto—and their satellites, all apparently uninhabited. And the search for exoplanets and planetary systems around nearby stars has yielded bounty beyond expectations. Techniques invented and honed in the past twenty years have allowed the discovery of these other worlds. Meanwhile, we have also listened carefully for potential radio signals from intelligent civilizations in a concerted fashion since the late 1970s.

The seemingly startling idea that we may not be alone in the universe, as we saw, originated well before the twentieth century; in fact, it dates back to ancient Greece. As I have recounted, the concept of other worlds and the possibility of extraterrestrial life transformed from heresy to orthodoxy between the fifth century BCE and the eighteenth century CE. The term *extraterrestrial life* is modern, but antecedents referred to the identical notion of the plurality of worlds.

Copernicus caused the dramatic shift in world view that shockingly displaced us from the center of the cosmos—imagine that! Galileo's telescopic observations of other planets in our solar system helped dismantle the Aristotelian notion of the uniqueness of the earth compared to all other celestial bodies. The atomist Epicurus had been on the right track after all, despite having been silenced by Plato and Aristotle for almost a millennium.

What motivates us to seek other habitable planets and evidence for intelligent life elsewhere? This quest began in earnest with the project SETI (Search for Extraterrestrial Intelligence) in the 1960s. Scientists working at the SETI Institute have since used microwave radio waves in an effort to detect potential communications from other solar systems. The first such targeted search for extraterrestrial life was conducted with the Green Bank Telescope, housed at the National Radio Astronomy Observatory in West Virginia. In 1960, a lanky twenty-nine-year-old postdoctoral fellow named Frank Drake began to wonder how he could exploit the newly built twenty-six-meter- (eighty-five-foot-) diameter radio dish antenna to try to catch celestial conversations from a planet orbiting a star about a dozen light-years away. At this point no one had detected a single exoplanet. There was no evidence suggesting that this would be a fruitful endeavor. Calculating the feasibility of such a detection, the ambitious Drake managed to persuade his supervisors that they ought to periodically point the telescope at two nearby sunlike stars, Tau Ceti and Epsilon Eridani, to listen in on any alien civilizations there that might be transmitting radio signals toward us on earth. Drake named this effort Project Ozma, after the princess who ruled over L. Frank Baum's fictional land of Oz. Although Project Ozma never detected anything other than interstellar static, it inspired an entire generation of scientists and engineers to take seriously the possibility of communicating with extraterrestrials. The 1982 Hollywood flick *E.T.*—a blockbuster and the highest grossing film of the 1980s—also whipped up and planted the idea of this possibility in the public imagination. The SETI project was funded to various degrees by the U.S. federal government until 1993. When this funding was cut, the institute registered as a nonprofit and began to operate solely on funds from private philanthropy. Paul Allen, one of the founders of Microsoft, funded an eponymous array, which now sits on the SETI campus in Mountain View, California. A more recent avatar, the project SETI@home has garnered wide public support. It is a scientific experiment that deploys the dormant power of Internet-connected computers in private homes. This was one of the first "citizen science" projects to crowd-source computer power from personal computers. You can participate by running a freely available program that you can install on your own personal computer at home to analyze radio telescope data procured by the SETI project whenever your computer is idle and not being used for other tasks.

Aside from ardently believing in the project of looking for alien life, Drake made a key contribution to the effort by quantitatively formulating the possibility of being contacted by or contacting life elsewhere. He did not wait for the future discoveries of exoplanets made by the Kepler satellite mission. This estimate, named the Drake equation, came out of a conference that he organized in 1961, whose sole purpose was to quantify whether SETI had any reasonable chance of detecting alien civilizations around other stars. Held in collaboration with the U.S. National Academy of Sciences, this informal meeting had interdisciplinary luminaries as attendees, including several Nobel laureates in chemistry and medicine and the physicist Philip Morrison, and perhaps the only undistinguished name at that time—a young postdoctoral fellow named Carl Sagan.

A few days prior to the meeting, while working on the agenda, Drake pondered the key ingredients—the crucial pieces of information that would be needed to estimate the number of detectable advanced civilizations that might currently exist in our galaxy. He began by assembling the probability of various factors, the first of which was the creation rate of habitable planets (the cosmic cradles for other civilizations), some fraction of which could host life and give birth to intelligent and sentient beings. Then he folded in the fraction of these civilizations that might have developed technologies and could communicate by beaming their signals across vast interstellar distances, factoring in the average longevity of such a society. The product of all these ifs, the many nested conditionals, was an estimate of the number of detectable advanced civilizations in the Milky Way. Via a series of intricate arguments combined with the sparse relevant data on the efficiency of star and planet formation that was then available, Drake realized that this number depends essentially only on one factor: the longevity of a technologically advanced civilization. The possibility of contact with extraterrestrials hinges on timing—on whether they are both alive and technologically advanced at the same time we are. Therefore, the more potentially habitable sites we can discover, the better the odds are that one of them may host a technologically advanced civilization. What is needed to start with is rapid discovery of as many exoplanets as possible, then to zero in on the ones that might be habitable and search for signatures of life. NASA's upcoming James Webb Space Telescope (JWST) mission, set to launch in 2018, will—among many other scientific goals—build on the Kepler satellite's legacy by helping to identify planets and planetary systems in the universe that may support life.

Simultaneously, the question of what even constitutes life needs to be addressed and understood. The debate is raging and will conceivably continue even if any traces are detected elsewhere. This is a tricky question, and the answer depends on where one is coming from in terms of disciplinary alignment. In his book *Life on Other Worlds,* the astronomer and historian of science Steven J. Dick chronicles this furious debate about the definition of life through the twentieth century. One of the articles that he discusses is by the evolutionary biologist George Gaylord Simpson, written in 1964 as the United States was preparing to step up the search for life on Mars. In this article, "The Nonprevalence of Humanoids," Simpson cautions that life elsewhere need not be similar to life on earth. The biologist Harold Blum had earlier labeled this point of view "opportunistic" and the opposing stance "deterministic." According to the deterministic argument, evolution unfolds in the same sequence everywhere in the universe, with complexity growing with time. The opportunistic view holds that life has many possible courses. Simpson noted that most exobiologists take the deterministic view, for which there is no empirical evidence, while evolutionary biologists tend to take the opportunistic one, which the fossil record on earth amply supports. Fossils reveal that most early life-forms went extinct and that evolutionary processes have had a considerable random component. Simpson argued that life, should it arise elsewhere in the universe, need not follow the arc that we are familiar with, from protozoa to human.[13]

A basic definition, on which many biologists agree, is that organisms capable of independent growth and replication are life-forms. But even this definition is not crystal clear and has gray areas, including whether viruses constitute life, given that they have their own genome but are not capable of reproducing on their own. This definition also seems to preclude prions, single rogue proteins that can replicate and are responsible for a host of ailments including bovine spongiform encephalopathy, or mad cow disease. Essentially, however, there is broad agreement that bacteria are the most rudimentary form of life.

If life itself is the subject of debate, the criteria conducive to life are also under scrutiny. Now that we are discovering exoplanets, there is a new urgency to understand what the necessary and sufficient conditions are for habitability—and how to detect any potential inhabitants. Originally, the mere presence of oxygen was believed to be sufficient, but now it is clear that many nonbiological processes can produce the elements thought necessary for life, and besides, for most of the time that life was evolving on earth, practically none of these gases were in the atmosphere. So oxygen and ozone that have been reliably detected in planetary atmospheres may not necessarily signal the presence of life teeming below on the surface. Meanwhile, arguments abound on whether other elements and compounds such as carbon dioxide, methane, and ammonia—seen as essential building blocks for life on earth—might be associated with the origin of life elsewhere. Closer to home, the issue of life is unsettled even on Mars, where we have not just been prospecting from afar, having successfully landed a rover that is analyzing soil samples with a mass spectrometer.

The search for exoplanets is continuing full tilt. The impressive current tally has demonstrated the efficacy of a couple of well-established detection techniques. The big open question is how many more earthlike planets are out there for us to uncover. Then, of course, comes the question of habitability and if there is life in any shape or form on any of these planets. Using data from NASA's Kepler satellite mission, a study in 2013 estimated that 22 percent of sunlike stars might harbor earthlike planets. Of course, some in the media ran amok with these numbers. Even the typically sober *New York Times* reported, "The known odds of something—or someone—living far, far away from earth improved beyond astronomers' boldest dreams on Monday." *USA Today* proclaimed, "We are not alone." A year earlier, the *Guardian* had run an article by Martin Rees titled "Are We Alone in the Universe? We'll Know Soon."[14]

How close are we actually to answering this question? The hype goes to the heart of what is called the Fermi paradox, named after the eminent Italian émigré physicist who was the father of controlled fission, which enabled the making of the atomic bomb in the United States during World War II. If we admit to not being exceptional as a life-form on earth, then there should be many other places in our galaxy and the universe where life exists. In that event we should have encountered some extraterrestrials by now; however, we have not. This is Fermi's paradox. In a 1975 paper published in the *Quarterly Journal of the Royal Astronomical Society,* Michael Hart addressed the issue of there being no intelligent beings from outer space on earth at the present time. From this fact, he concluded that there is "strong evidence that we are the first civilization in our Galaxy."[15] His argument rests on time scales—if it exists, intelligent life everywhere in our galaxy should take several million years to prosper but then billions of years to evolve. So logically, if extraterrestrial life existed, then our alien compatriots would be here already, and since they are not here, they don't exist. As with the Drake equation estimates, it is really all down to timing.

There is a sharp divide in attitude, world view, and expectation between many biologists and many astronomers that can be pegged to their respective perspectives on whether the earth is unique. Astronomers, most notably Sagan and Drake, typically believe that we are not unique or particularly special in any way. Sagan thought that the unique earth hypothesis was flawed. He resolved the Fermi paradox by inferring that all extraterrestrial civilizations that have decided to reach out are doing so very slowly and therefore have not made contact with us yet. Biologists, on the other hand, generally believe in the uniqueness of life on earth; the richness and complexity that they constantly encounter make them confident that our planet is very special, in that it is the only one that harbors intelligent life, although this development involved randomness. Stephen Jay Gould has famously said that if the tape of life and evolution were to rewind and run again, it's totally unclear whether humans would be an end product. In fact, the evolutionary biologist Theodosius Dobzhansky noted that of the more than two million species that call earth home, only one has evolved language, created and transmitted culture, and has awareness of self, life, and death. He therefore considered it absurd to hold the opinion that if life exists anywhere else it must eventually give rise to rational beings.[16]

Sagan's extreme optimism about intelligent extraterrestrial life stemmed from his belief that the utter mediocrity of our position in space and time makes us ordinary in the universe. Furthermore, he saw this discussion as a reflection of the age-old anthropocentric world view, dating back at least to Claudius Ptolemy. There is a real gap in how astronomers and biologists perceive our significance as a species and as intelligent life on this one planet. These two outlooks are informed by the intellectual take of their particular professions and are each justified in their own way.

The debate about our uniqueness has occasionally grown heated, not because of any new discoveries that challenge our world view quite yet but because of politics—which has partly revolved around the issue of how and if research dollars from the U.S. government should fund the SETI enterprise in the long term. It is the divergence of opinion on the probability of detecting extraterrestrial life that ultimately led to SETI's loss of its governmental funding.

But SETI was far from the first organized search for intelligent life. A mere hundred years after Copernicus dislodged earth from its position as the universal pivot, Rene Descartes suggested that the sun was not unique either. In his 1644 *Principles of Philosophy,* he wrote that all other stars in the sky were akin to our sun. Each of these, he postulated, could harbor its very own posse of planets, perhaps even with inhabitants with souls—therefore infinitely many planets. His speculation that other stars could host planets had to wait for confirmation for more than 350 years.

In 1995, two Swiss astronomers, Michel Mayor and Didier Queloz, accidently discovered the first planet outside our solar system, around the star 51 Pegasi, a mere fifty-one light-years away. This planet, circling around its star every four days or so, is six times closer to its sun than Mercury is to ours. It is also incredibly massive, weighing in at almost half as much as Jupiter, the behemoth of our solar system's planetary lineup. This planet around 51 Pegasi, it turns out, was the first in a class of exoplanets referred to as hot Jupiters—massive planets precariously and dangerously close to their host stars.

Any planet on its own does not emit light to be detectable in comparison to the star it circles. It merely reflects light emanating from its parent star. Aside from the challenge of detecting such a faint light source on its own, the bright light from the parent star obscures and further impairs detection. For these reasons, astronomers have observed few of these planets directly and are rarely able to distinguish the planet from its host star.

Instead, they have generally resorted to indirect methods to detect exoplanets, again relying on gravitational effects, as is necessary for detecting dark matter and black holes. Several such methods have yielded success. The most popular search strategy, which revealed the hot Jupiters, is to measure radial velocity—the wobble of stars caused by the pull of their planet companions. Observation from ground-based telescopes then confirms the planet's existence. This is how the method works: A star harboring a planet will respond to the planet's gravity with motion around a tiny orbit. This leads to detectable small changes in the speed of the star that we can measure—manifested as a small change in its radial velocity with respect to the earth. This change in radial velocity can be inferred from the Doppler shift produced in the star's spectrum. This method is distance independent, but finding lower-mass planets that produce tiny wobbles requires high-fidelity data, therefore this method is generally used only nearby stars, where nearby is about 160 light-years or less from earth. It is impossible to track many target stars at the same time with a single telescope. Jupiter-size planets can, however, be detected up to a few thousand light-years away. This method preferentially detects massive planets that are close to their stars. It is a lot easier to detect planets around lower-mass host stars. Because the gravitational effect from their planets can be more easily discerned, and besides low-mass stars generally rotate more slowly. Fast rotations smudge spectral lines, making detection harder. The mass of the planet can be directly inferred from these radial velocity measurements. Hot Jupiters are the easiest to detect via this wobble, and it is therefore no surprise that all early detections with radial velocity measurements—including the first, by Mayor and Queloz—yielded this varietal. Hot Jupiters are not expected to form right where they are seen, so close to the parent star, but rather at large distances, and are then believed to migrate inward, toward the star. These hot, puffed-up gas giants are typically hotter than Venus in our solar system and are as inhospitable as it gets to life as we know it.

With improved instrumentation, astronomers have recorded tinier and tinier wobbles and so have gradually discovered less massive planets. Initially the two main competing groups in the exoplanet hunting mission were the Swiss duo and a group at the University of California, Berkeley, led by Geoffrey Marcy. Both have been monitoring nearby stars and making velocity measurements for close to two decades. One of the first planet-hunting honors that the Berkeley group had was the discovery of the first multiplanetary system, around the star Upsilon Andromedae, about forty-four light-years away. It currently appears to be four planets, orbiting a system of two stars. All of the planets seem comparable in size to Jupiter.

The second method for detecting planets is the transit method, and this is what NASA's Kepler spacecraft used to find thousands of candidates.[17] If a planet transits across the face of its parent star, then the measured brightness of the star will diminish by a very small amount. We recently saw this in our solar system with the transit

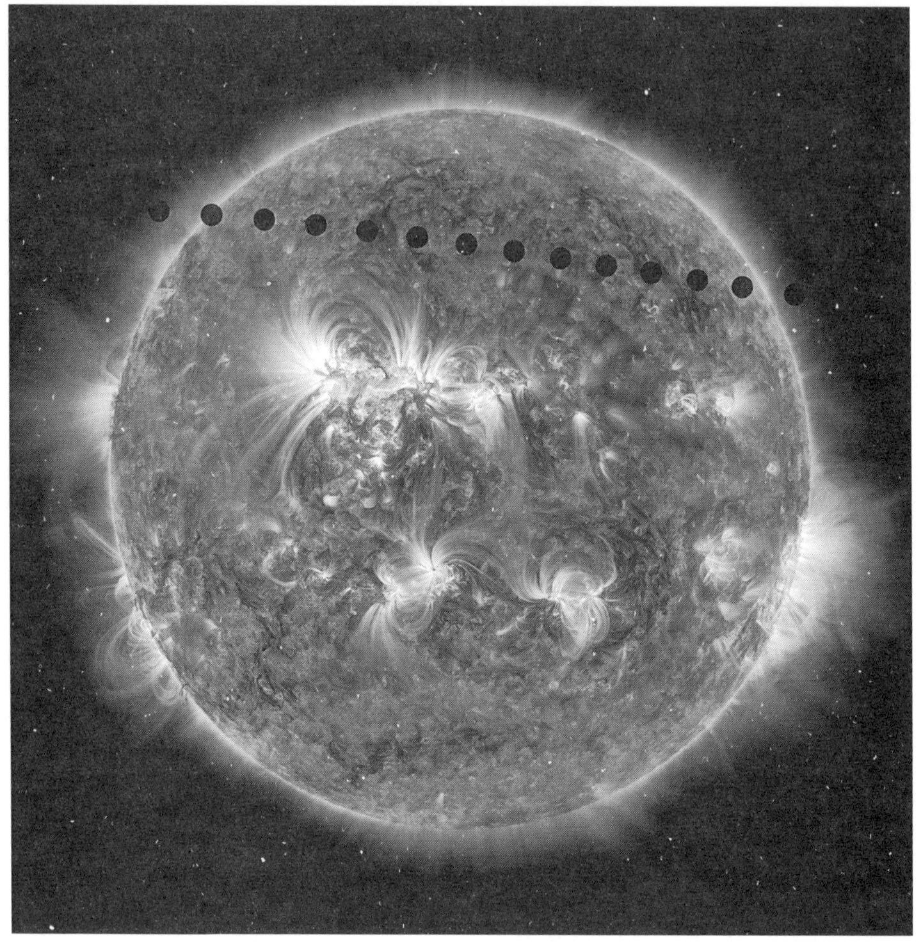

The transit of Venus in 2012. Compiled sequence of 171 images, courtesy of NASA/Goddard Space Flight Center/SDO.

of Venus across the face of the sun in 2012. This diminution can be measured and is used to determine the radius of the planet.

The amount of this dimming of course depends on how much larger the star is compared to the size of the planet. For the star HD 209458, the dimming during the transit was less than 2 percent. This method has the drawback that transits are visible only for specific alignments of the planet-star orbit with respect to the observer. Besides, they can be seen only from space, because of the distorting effect of our atmosphere. Fortunately, the imaging camera aboard the Kepler spacecraft has the precision required to measure the dimming of light from the star during the transit at the level of a few percent.

As of November 10, 2015, more than 5,000 exoplanets have been discovered, some of them in the 484 known multiplanetary systems. Beyond our solar system, the stars with the most confirmed exoplanets are Kepler-90 and HD 10180, with seven and nine, respectively. The star Gliese 876, with four confirmed exoplanets, is the closest multiplanetary system to us, at fifteen light-years away. Including these four, there are twelve multiple-planet systems within fifty light-years of us, but most are much farther away.

Now planet hunting is a veritable industry. In fifteen years it moved from a small group effort involving a handful of observers and instrumentation builders to big science, with the development and deployment of the Kepler satellite, whose scientific team numbered in the hundreds. Seasoned planet hunters note that because essentially all stars have accompanying planets and some host multiplanetary systems, a simple estimate suggests that the number of planets in the visible universe must exceed the number of stars! That's extremely encouraging for the prospects of detecting life, because it is expected to emerge on planets or their moons. For carbon-based life, a certain temperature range is required, because otherwise organic molecules disintegrate. In the extremely cold temperatures expected in outer space, the rates of chemical reactions slow considerably, and that makes the rapid metabolic reactions required for the existence of intelligent creatures very difficult or even impossible. So it appears there is a sweet spot for carbon-based life—only planets and moons with moderate temperatures might be able to host sophisticated and advanced life-forms that are familiar and similar to us.

To be honest, as with the anthropic principle discussed previously, I find that earthlings take a rather self-centered view of this subject. The arguments and estimates I've discussed hinge on a completely anthropocentric view of life. Is it a lack of imagination that constrains us from wandering beyond life as we know it or intelligence as we see it manifested? It is entirely possible that intelligent life might arise in a range of less familiar environments by pathways unknown to us. The detection of extremophiles on earth gives us an inkling of the kinds of life-forms that could stay alive over long time scales in the harshest conditions. Examples include bacteria that thrive at extremely high temperatures, such as *Thermus aquaticus* (which enjoys a balmy existence at 71 degrees Celsius—160 degrees Fahrenheit—in hot springs in Yellowstone National Park) and *Pyrolobus fumarii* (which was discovered in the smoker vents in the Mid-Atlantic Ridge and lives at 113 degrees Celsius, or 235 degrees Fahrenheit), or extremely low temperatures, including cryophiles such as *Psychrobacter*, which grow between −10 and 42 degrees Celsius (14 and 108 degrees Fahrenheit). These organisms reveal that life is adaptable and durable despite the violent changes on earth such as its five known ice ages and the reversal of its magnetic fields more than 11,700 years ago. So who knows what kinds of life-forms can thrive on exoplanets? Many of these exoplanets are far enough away from their parent stars that water, if it were there, would remain liquid—so many could in fact be super-Earth worlds (planets that are slightly more massive than Earth but less so than Uranus or Neptune). In an exciting development, in July 2015 Stephen Hawking and the Russian billionaire Yuri Milner held a press conference in London to announce their new joint project, Breakthrough Listen, the largest initiative yet to detect signals of alien life. Milner's generous philanthropic gift of one hundred million dollars is to be used partly to buy dedicated time on two existing, far-flung radio telescopes that will then search in concert for extraterrestrial signals: the Robert C. Byrd Green Bank Telescope in Virginia and the Parkes Radio Telescope in New South Wales, Australia. Once again, like the Event Horizon Telescope, which is configured to detect the black hole at the center of our galaxy, this setup transforms the entire earth into one large radio telescope with an enormous baseline, in this case equivalent to a single dish that stretches from the United States to Australia for listening to radio signals transmitted by intelligent beings elsewhere. Announcing this collaborative effort, Hawking said, "In an infinite universe, there must be other occurrences of life. . . . Or do our lights wander a lifeless cosmos[?] . . . Either way, there is no bigger question." As for Milner, he revealed that his motivation for this initiative arose from his gut feeling that we're not alone. Speaking at the press conference, he noted, "I think it's a low probability but high impact event. Irrespective of what the answer is, it's a powerful answer. At any given time, we should apply the best technology and use the best instruments available to search for that answer."[18] We eagerly await eavesdropping on alien conversations.

Central to all these debates—on what defines life and intelligent life, on what life at other locations in the universe might look like, and on how we might best make contact with other sentient beings if they exist—is the issue of how unique we really are. What is ultimately at stake is defining and finding our place. The dramatic progress in cosmology to date has revealed that we are not special. Our planet is not the center of the universe. Our solar system, and potentially our universe, is simply one of many. Our universe is also a rather peculiar one, whose principal ingredients, dark matter and dark energy, remain elusive. Our eyes are not tuned to see the majority of reality. Yet although we are seemingly insignificant, we have significant capabilities as a species. We have tackled and answered questions about the universe that once seemed impossible and intractable. We have made remarkable progress in what we know and how we have come to know it. Despite the limit of our cognitive capacities, with brains that fit into a skull the size of a melon, we have deciphered many secrets of the wondrous universe with our explorations. Yet we are also at the brink of destroying the intelligent life that we know exists and the verdant planet that has nurtured us. Along with cosmic curiosity, we have urgent terrestrial responsibilities, ones that we cannot ignore.

We now wait with anticipation and excitement for the detection of the whiff of an organic molecule, the fingerprint of a biomarker, or evidence of water from the recently discovered other worlds. Intelligent life seems remote, but we've already come so far: from John Michell, who never imagined that we could trace the orbit of stars around the Milky Way's central black hole, to the technologies that we have developed today, which were beyond the wildest imagination of scientists even fifty years ago.

So what's left to map and understand? The current frontier spans both the largest scales and the smallest. On one extreme, we look beyond our universe. On the other, we look carefully inside our galaxy and reflect on ourselves. In both pursuits we seek companions—a menagerie of universes, with all manifestations of the six fundamental constants, and a community of intelligent beings, which force us to question what it means to be alive.

EPILOGUE

> Bold ideas, unjustified anticipations, and speculative thought are our only means for interpreting nature. . . . Those among us who are unwilling to expose their ideas to the hazard of refutation do not take part in the scientific game.
>
> —Karl Popper, *The Logic of Scientific Discovery*

In this book I have traced the journey of radical scientific ideas from opposition to acceptance. You have seen that resistance can be complicated and not entirely intellectual. Personal rivalries, fame, and dogma can keep the scientific community from reaching consensus. But we have also seen that the power of persuasion ultimately lies with data and evidence, preferably collected from many independent lines of inquiry.

I hope I have also shown that the most interesting scientific ideas are types of maps—some literal. With each revision, we see the details of the world around us more clearly. We abandon places that proved realms of fantasy. And, if we are very fortunate, we spot terra incognita, mysterious lands for us to explore further. To advance, we need human ingenuity, scientific minds that are flexible and open to change but also exacting—persuaded only when the evidence is overwhelming. We also need technology, the engineering of increasingly sophisticated instruments that enable more accurate measurements.

This marriage of technology and ingenuity has spearheaded a fundamental shift in the scale and practice of the research enterprise. The past forty years have witnessed the transition to big science—large projects that involve huge investments of economic, human, technical, and intellectual capital. Brilliant individuals working in isolation can no longer tackle many of the pressing questions at the cutting edge of cosmology. Current challenges require massive teams of hundreds of bright scientists contributing specialized training to nearly corporate endeavors. This shift in style is not superficial; it marks a significant change in culture and intellectual approach. We now require big, expensive pieces of equipment and great resources—large telescopes on the ground, space telescopes, and powerful supercomputers. While the majority of the scientific community has celebrated this new wave of big projects, there are also debates about how this transformation has affected the role of the individual creative scientist, driven by personal curiosity. The question centers on intellectual risk. Individuals can take certain risks that groups, which require a level of consensus from the start to obtain funding and set research directions, cannot. We must be careful that we do not limit ourselves to groupthink, as explorers cannot always be safe, cautious, and conservative. The solution is to nurture and promote both research styles—tapping into the power of individual visionaries and harnessing the efficiency and rapid pace that collaborative groups afford.

Along with big science has come another major cultural shift: open access—catalyzed by the Internet's ability to disseminate information cheaply and quickly. This new culture and technology enabled the creation of places like arXiv.org, where astronomers and astrophysicists post their papers—often as soon as they submit their manuscripts to journals for peer review. This openness has vastly improved access and proffered a new way for scientists to lay claim to their work that is time-stamped. Physicists pioneered this way of sharing their ideas, but with the introduction and mass appeal of social media, of course now everyone is in on the game. In turn, social media has allowed scientists to directly court journalists and the public as a means of sharing their exciting new results with a wider audience.

The pressures of funding big science have contributed to this desire for transparency. If large amounts of public taxpayer money support multimillion-dollar state-of-the-art equipment, then scientists must not only justify what they do but also share their results as broadly and quickly as possible. This brisk pace has contributed to the golden age of cosmology, and we have seen spectacular breakthroughs.

But beyond allowing access, the Internet has opened science itself, enabling the public to both watch and participate in debates and discussions. With the help of instantaneous global media, the public can now view the process of how researchers reach scientific consensus in real time as they settle matters. Take, for example, the claimed detection of gravitational waves in 2014. This was a huge headline, and rightfully so; the hunt for gravitational waves, a consequence of Einstein's theory of general relativity, began in 1916. Physicists from the BICEP2 team operating a telescope at the South Pole claimed evidence for these waves and therefore for cosmic inflation. This would constitute incontrovertible evidence for the big bang model of the origin of the universe. The BICEP2 team addressed a press conference prior to peer review of its results. The news hit the science blogs and went viral. And the rest of the cosmology community jumped in to understand and replicate the results. After much debate among experts, the potential gravitational waves appeared to be only cosmic dust.

This is an interesting case study of a high-impact research result that spread instantaneously in a globally connected world and eventually did not hold up to detailed scientific scrutiny—a warning that we should be very careful in analyzing and sharing our data or risk embarrassment. Many scientists were extremely unhappy with this turn of events. I am not one of them, although I do think that peer review is an important and integral part of professional science that should not be circumvented. I believe open science is both useful and crucial for the public to understand what we do. In fact, now scientists have no choice but to engage with the public, share their data and analysis with professional collaborators as well as amateur researchers around the globe. Given all this, why are we still witnessing the most vehement denial of science? In my opinion, what fuels rampant denialism is not lack of knowledge of scientific *facts* but rather ignorance about how science and scientific reasoning *work*. Pulling back the curtain on the scientific process for the public to watch and understand will quell the disbelief. I also like to think that current scientific skepticism is in part a consequence of hysteresis—that lag that makes it difficult to cope with the rapid and unpredictable nature of scientific discovery and transformation in the digital age.

If this is a book about maps, it is equally one about time and place. We are living in a disorienting universe, whose expansion is accelerating. And at no other period in human history have we had to contend head on so frequently with the provisionality of our understanding. We have a cosmic map that is eternally in flux. The fact that by their nature scientific truths are subject to refinement and revision is now an inescapable part of our reality. Our world view has shifted sharply in the past hundred years, rewriting the very sense of who we are, where we came from, and where we are headed.

Now you know that our perspective has come in part from looking outward. But it also comes, more intimately, from the other direction. Like our universe, our brains and genes were once thought fixed. We now understand that they are mutable. The mapping of the genome has helped us to uncover the cartography of our chemical essence. Developments in computational genomics in the past decades have helped establish Africa as the cradle of all humanity and have traced our subsequent migration. Tremendous progress in neuroscience is now enabling us to unravel how the human brain works. Functional MRI (magnetic resonance imaging) offers an unprecedented and noninvasive view. We have uncovered new interconnections between neurons and how they function in unison. That said, the way that dynamic and ever-reconfiguring networks cohere and synchronize is largely uncharted territory that awaits new voyages of exploration. Human brains and genes are complex, but there is hope that we may soon figure out how the flipping of a particular neuronal switch—or group of switches—can jump-start our thinking. The result: a map of an emotion, a movement, or even a new idea. Maps continue to shape our view of the cosmos and ourselves.

NOTES

1. Early Cosmic Maps

1. A detailed description of Layard's discoveries in his own words can be found in Austen Henry Layard, *A Popular Account of Discoveries at Nineveh* (New York: Derby, 1854). Incidentally, forty thousand of these cuneiform tablets are preserved in Yale University's Babylonian Collection—the world's fourth largest collection of ancient Mesopotamian cuneiform tablets and the single largest one held outside a museum.

2. Kenneth R. Lang, *The Cambridge Guide to the Solar System* (Cambridge: London, 2011), 410–20. A more detailed account of Babylonian astronomy, astrology, and cosmological speculation can be found in Thorkild Jacobsen, "Enuma Elish—'The Babylonian Genesis,'" in *Theories of the Universe: From Babylonian Myth to Modern Science,* ed. Milton K. Munitz (New York: Free Press, 1965), 8–21. The history of astronomy richly documents the connections drawn between the heavens and the earth as portrayed in maps. For more elaborate examples see my review essay "Revelations from Outer Space," *New York Review of Books,* May 21, 2015, www.nybooks.com/articles/archives/2015/may/21/interstellar-revelations-outer-space/.

3. A translation of the Venus tablet appears in Nicholas Campion, "Astrology in Babylonia," in *Encyclopaedia of the History of Science, Technology, and Medicine in Non-Western Cultures,* 2nd ed., ed. Helaine Selin (Berlin: Springer Verlag, 2002), 251.

4. Carlo Rovelli, *The First Scientist: Anaximander and His Legacy* (Yardley, PA: Westholme, 2011), 57–60, 104. Additional information on the Milesians can be found in F. M. Cornford, "Pattern of Ionian Cosmogony," in Munitz, *Theories of the Universe,* 21–31; and G. S. Kirk, J. E. Raven, and M. Schofield, eds., *The Presocratic Philosophers: A Critical History with a Selection of Texts,* 2nd ed. (Cambridge: Cambridge University Press, 1983), 76–142.

5. John Vardalas, "A History of the Magnetic Compass," *Institute,* member

newspaper of the Institute of Electrical and Electronics Engineers, November 8, 2013, http://theinstitute.ieee.org/technology-focus/technology-history/a-history-of-the-magnetic-compass. John Huth's recent *The Lost Art of Finding Our Way* (Cambridge, MA: Harvard University Press, 2013) is a history of navigation that includes a discussion of primitive techniques used by ancient cultures as well as a how-to of the modern age.

6. R. C. Taliaferro provides a translation of the *Almagest* in *Great Books of the Western World,* vol. 16 (Chicago: Encyclopaedia Britannica, 1952). For reviews of ancient Greek cosmologies that predate Ptolemy, see the translations with commentary of F. M. Cornford, *Plato's Cosmology: The Timaeus of Plato* (New York: Humanities Press, 1937); and of W. K. C. Guthrie, *Aristotle: On the Heavens,* Loeb Classical Library (Cambridge, MA: Harvard University Press, 1939).

7. A more detailed description of Islamic science in the Middle Ages, with a focus on mathematics and its application to astronomy, can be found in Ali Abdullah al-Daffa, *The Muslim Contribution to Mathematics* (London: Croom, 1977).

8. Petra G. Schmidl, "Two Early Arabic Sources on the Magnetic Compass," *Journal of Arabic and Islamic Studies* 1 (1997–98): 85.

9. The cosmic views depicted in the *Breviari* image, as well as others that this chapter subsequently discusses—namely, those from the *Catalan Atlas* and by Andreas Cellarius, Giovanni Battista Riccioli, and Bernard Picart—are included in Michael Benson's comprehensive compendium of images *Cosmigraphics: Picturing Space Through Time* (New York: Abrams, 2014), which I reviewed in "Revelations from Outer Space."

10. Owen Gingerich, *The Book Nobody Read* (New York: Penguin, 2005), 146.

11. In fact, Aristarchus of Samos (310–230 BCE) is credited with having first proposed a heliocentric model for the solar system. Although the text in which he advanced it does not survive, subsequent reference to his mathematical calculations appears in a book by Archimedes (287–212 BCE).

12. Gingerich, *Book Nobody Read,* 170–85.

13. Details of the fruitful scientific collaboration between Brahe and Kepler can be found in Kitty Ferguson's *Tycho and Kepler: The Unlikely Partnership That Forever Changed Our Understanding of the Heavens* (New York: Walker, 2002).

2. The Growing Border

1. Walter B. Clausen, Associated Press release, February 4, 1931, quoted in Gale E. Christianson, *Edwin Hubble: Mariner of the Nebulae* (Chicago, University of Chicago Press, 1995), 210.

2. Jordan Holliday, "Before Revolutionizing Astronomy, Hubble Helped Rewrite Record Books," *Chicago Maroon,* April 10, 2009, http://chicagomaroon.com/2009/04/10/before-revolutionizing-astronomy-hubble-helped-rewrite-record-books/; Alan Lightman, *The Discoveries: Great Breakthroughs in 20th-Century Science, Including the Original Papers* (New York: Pantheon, 2005), 230; Marcia Bartusiak, *The Day We Found the Universe* (New York: Vintage, 2009), 170; and "Rhodes Scholars: Complete List, 1903–2015," www.rhodeshouse.ox.ac.uk/about/rhodes-scholars/rhodes-scholars-complete-list. Lightman and Bartusiak offer finely researched, in-depth biographical accounts of Hubble; in addition to their titles above, see Bartusiak, *Archives of the Universe: 100 Discoveries That Transformed Our Understanding of the Cosmos* (New York: Vintage, 2004), 414–24.

3. Bartusiak, *Day We Found the Universe,* 174; and Christianson, *Edwin Hubble,* 86–87.

4. Aristotle, *On the Heavens,* book 1, chapter 3, translated by W. K. C. Guthrie, Loeb Classical Library (Cambridge, MA: Harvard University Press, 1971), 25. Fixed stars do have parallax, which is a change in apparent position caused by the orbital motion of the Earth. This effect is small enough not to have been noticed until modern times. It can be used to find the distance to nearby stars.

5. *Liber Hermetis, Part I,* translated by Robert Zoller, edited by Robert Hand (Berkeley Springs, WV: Golden Hind, 1993). The *Almagest* is Ptolemy's second-century BCE treatise on astronomy and mathematics. The geocentric view it presents held sway for more than twelve hundred years, until Copernicus, making this one of the most influential texts of all time.

6. William Shakespeare, "Sonnet 21," in *Sonnets,* ed. Thomas Tyler (London: D. Nutt, 1890), available at *Shakespeare Online,* accessed August 12, 2014, www.shakespeare-online.com/sonnets/21.html.

7. Percy Bysshe Shelley, *Queen Mab; a Philosophical Poem* (New York: William Baldwin, 1821), 46.

8. Robert Mitchell, "'Here Is Thy Fitting Temple': Science, Technology and Fiction in Shelley's *Queen Mab,*" in "Romanticism on the Net," special issue, *Romanticism and Science Fiction* 21 (February 2001): www.erudit.org/revue/ron/2001/v/n21/005964ar.html.

9. A. Einstein, "Kosmologische Betrachtungen zur allgemeinen Relativitatstheorie" [On the cosmological problem of the general theory of relativity], *Sitzungsberichte der Preussischen Akademie der Wissenschaften* 1 (1917): 142–52, translated by W. Perrett and G. B. Jeffrey as "Cosmological Considerations of the General Theory of Relativity" in H. A. Lorentz, Einstein,

H. Minkowski, and H. Weyl, *The Principle of Relativity* (New York: Dover, 1952), 175–88.

10. A. Einstein, "Cosmological Considerations," 188. Einstein apparently later claimed that this mathematical term, the cosmological constant, was his "biggest blunder." The precise attribution is somewhat cloudy. See Mario Livio, *Brilliant Blunders: From Darwin to Einstein—Colossal Mistakes by Great Scientists That Changed Our Understanding of Life and the Universe* (New York: Simon and Schuster, 2013), 233.

11. Lightman, *Discoveries,* 230–32.

12. Georges Lemaître, "Un univers homogène de masse constante et de rayon croissant rendant compte de la vitesse radiale des nébuleuses extragalactiques," *Annales de la Société scientifique de Bruxelles* 47A (1927): 49–59, translated as "A Homogeneous Universe of Constant Mass and Increasing Radius Accounting for the Radial Velocity of Extra-galactic Nebulae," *Monthly Notices of the Astronomical Society* 91 (1931): 483–90, quote on 489.

13. Charles Darwin to W. D. Fox, February 15, 1851, available at *Darwin Correspondence Project,* accessed August 12, 2014, www.darwinproject.ac.uk/entry-94.

14. Immanuel Kant, *Allgemeine Naturgeschichte und Theorie des Himmels* (Königsberg: Petersen, 1755), part 1, translated by Hubble himself in his *Realm of the Nebulae* (New Haven: Yale University Press, 1982), 23–25.

15. Hubble, *Realm of the Nebulae,* 23.

16. Barbara L. Welther, "Pickering's Harem," *Isis* 73 (1982): 94.

17. Lightman, *Discoveries,* 111–26; and Henrietta Leavitt, "Periods of 25 Variable Stars in the Small Magellanic Cloud," article signed by Edward C. Pickering, *Harvard College Observatory Circular* 173 (March 3, 1912): 3.

18. Harlow Shapley, "Globular Clusters and the Structure of the Galactic System," *Publications of the Astronomical Society of the Pacific* 30, no. 173 (1918): 42–54.

19. The westward shift of astronomy is described in more detail in Robert W. Smith, "Edwin P. Hubble and the Transformation of Cosmology," *Physics Today,* April 1990, 52–58.

20. A detailed account of Hubble and his dealings—competitive and collaborative—with his scientific colleagues can be found in Bartusiak, *Day We Found the Universe,* 169–250; and Lightman, *Discoveries,* 236–40. Hubble's account of his research work is in his *Realm of the Nebulae.*

21. A light-year is the distance that light travels in a year—given that its speed is three hundred thousand kilometers (186,000 miles) per second, this corresponds to a distance of about 9.7 trillion kilometers (six trillion miles).

22. Hubble, *Realm of the Nebulae,* 23.

23. Smith, "Edwin P. Hubble," 57.

24. George W. Gray, "Invisible Stuff," *Atlantic Monthly,* July 1931, 47–56.

25. James Stokley, "Eddington Pictures Expanding Universe," *New York Times,* September 7, 1932, available at http://timesmachine.nytimes.com/timesmachine/1932/09/08/100802822.html.

26. Edwin Hubble, "Effects of Red Shifts on the Distribution of Nebulae," *Astrophysical Journal* 84 (1936): 517–54, quote on 517.

27. Cormac O'Raifeartaigh, Brendan McCann, Werner Nahm, and Simon Mitton, "Einstein's Steady State Theory: An Abandoned Model of the Cosmos," accepted for publication in *European Physics Journal H,* last revised May 22, 2014, http://arxiv.org/abs/1402.0132.

28. Fred Hoyle, BBC radio broadcast, March 28, 1949, reprinted in *Listener* 41 (April 7, 1949): 568. See www.joh.cam.ac.uk/library/special_collections/hoyle/exhibition/radio.

29. Fred Hoyle, "Steady State Cosmology Revisited," in *Cosmology and Astrophysics: Essays in Honor of Thomas Gold,* ed. Yervant Terzian and Elizabeth M. Bilson (Ithaca: Cornell University Press, 1982), 51.

30. Helge Kragh, *Cosmology and Controversy: The Historical Development of Two Theories of the Universe* (Princeton: Princeton University Press, 1996).

3. The Dark Center

1. Rumiko Takahashi, *Inuyasha,* originally serialized in *Weekly Shōnen Sunday* from 1996 to 2008. Miroku also appears in season 1, episode 16 of the *Inuyasha* anime, which first aired on February 19, 2001. See also Rupert W. Anderson, *The Cosmic Compendium: Black Holes* (self-published through Lulu .com, 2015), 57.

2. Stanley Wolpert, *A New History of India,* 8th ed. (New York: Oxford University Press, 2009), 185; J. H. Little, "The Black Hole—the Question of Holwell's Veracity," in *Bengal, Past and Present: Journal of the Calcutta Historical Society* 12 (1916), part 1, serial 23: 32–42, 136–71; and "Only One Topic in Paris," *New York Times,* May 29, 1887, accessed September 15, 2015, http://query.nytimes.com/mem/archive-free/pdf?res=9C04E0DB1730E633A2575AC2A9639C94669FD7CF.

3. Edgar A. Poe, "The Premature Burial," *Philadelphia Dollar Newspaper,* July 31, 1844, available at www.eapoe.org/works/info/pt048.htm#text02.

4. Bartusiak explores this question in detail in *Black Hole: How an Idea*

Abandoned by Newtonians, Hated by Einstein, and Gambled On by Hawking Became Loved (New Haven: Yale University Press, 2015), chapter 7.

5. Michael Sanderson, *Education, Economic Change and Society in England 1780–1870* (Cambridge: Cambridge University Press, 1995), 40. The Thirty-Nine Articles of Religion, which were finalized in 1571 and incorporated into and disseminated via the *Book of Common Prayer,* constitute the post-Reformation doctrine of the Church of England.

6. "Case Study: John Michell and Black Holes," excerpt from *Cosmic Horizons: Astronomy at the Cutting Edge,* ed. Steven Soter and Neil deGrasse Tyson (New York: New Press, 2000), on the American Museum of Natural History website, www.amnh.org/education/resources/rfl/web/essaybooks/cosmic/cs_michell.html.

7. John Michell, "On the Means of Discovering the Distance, Magnitude, andc. of the Fixed Stars, in Consequence of the Diminution of the Velocity of Their Light, in Case Such a Diminution Should Be Found to Take Place in Any of Them, and Such Other Data Should Be Procured from Observations, as Would Be Farther Necessary for That Purpose. By the Rev. John Michell, B. D. F. R. S. In a Letter to Henry Cavendish, Esq. F. R. S. and A. S.," *Philosophical Transactions of the Royal Society of London* 74 (1784): 35–57; and Charles Coulston Gillispie, *Pierre-Simon Laplace, 1749–1827: A Life in Exact Science,* with the collaboration of Robert Fox and Ivor Grattan-Guinness (Princeton: Princeton University Press, 2002), 175.

8. A. Einstein, "Ist die Trägheit eines Körpers von seinem Energieinhalt abhängig?" [Does the inertia of a body depend on its energy content?], *Annalen der Physik,* 4th ser., vol. 18 (1905): 639–41; and Einstein, "Erklärung der Perihelbewegung des Merkur aus der allgemeinen Relativitätstheorie" [Explanation of the perihelion motion of Mercury from the general theory of relativity], *Sitzungsberichte der Königlich Preussischen Akademie der Wissenschaften* 2 (November 18, 1915): 831–39.

9. Mitchell Begelman and Martin Rees, *Gravity's Fatal Attraction: Black Holes in the Universe,* 2nd ed. (Cambridge: Cambridge University Press, 2005), is a comprehensive introduction to black holes—their origin, peculiar properties, and cosmic roles—with copious illustrations and diagrams. Black holes in excess of one million times the mass of our sun are referred to as supermassive.

10. Einstein, "Trägheit eines Körpers."

11. Michael White and John Gribbin, *Einstein: A Life in Science* (London: Simon and Schuster, 1993), 115–16; and Frank Watson Dyson, Arthur Stanley Eddington, and Charles Davidson, "A Determination of the Deflection of Light by the Sun's Gravitational Field, from Observations Made at

the Total Eclipse of 29 May 1919," *Philosophical Transactions of the Royal Society* 220A (1920): 291–333.

12. A. Einstein, *Relativity: The Special and General Theory,* translated by Robert W. Lawson, introduction by Roger Penrose (New York: Pi, 2005); and Mario Livio, *Brilliant Blunders: From Darwin to Einstein—Colossal Mistakes by Great Scientists That Changed Our Understanding of Life and the Universe* (New York: Simon and Schuster, 2013), 269. I discuss more generally the provisional nature of science, for instance how new theories displace older ones, in my review essay "What Scientists Really Do," *New York Review of Books,* October 23, 2014, www.nybooks.com/articles/archives/2014/oct/23/what-scientists-really-do/.

13. Karl Schwarzschild, "Über das Gravitationsfeld eines Massenpunktes nach der Einsteinschen Theorie," *Sitzungsberichte der Königlich Preussischen Akademie der Wissenschaften* 7 (1916): 189–96, translated by Salvatore Antoci and Angelo Loinger as "On the Gravitational Field of a Mass Point According to Einstein's Theory," submitted May 12, 1999, http://arxiv.org/abs/physics/9905030; and Roy P. Kerr, "Gravitational Field of a Spinning Mass as an Example of Algebraically Special Metrics," *Physical Review Letters* 11, no. 5 (1963): 237.

14. The conflict between Chandra and Eddington is described in Kameshwar C. Wali, *Chandra: A Biography of S. Chandrasekhar* (Chicago: University of Chicago Press, 1992), 123–46. Arthur I. Miller's *Empire of the Stars: Obsession, Friendship, and Betrayal in the Quest for Black Holes* (Boston: Houghton Mifflin, 2005) is entirely about the controversy surrounding black holes.

15. Miller, *Empire of the Stars,* 3–15, 96–119, 135.

16. Ibid., 125–50.

17. Max Planck, *Scientific Autobiography and Other Papers,* translated by F. Gaynor (New York: Philosophical Library, 1949), 33–34.

18. A nanometer is one-billionth of a meter (1 nanometer = 10^{-9} meters).

19. "Riccardo Giacconi—Facts," fact sheet for the Nobel Prize in Physics 2002, www.nobelprize.org/nobel_prizes/physics/laureates/2002/giacconi-facts.html.

4. The Invisible Grid

1. Richard Panek, *The 4% Universe: Dark Matter, Dark Energy, and the Race to Discover the Rest of Reality* (New York: Mariner Books, 2011), chapters 1–6.

2. For Fraunhofer's biography see T. Hockey, ed., *The Biographical Encyclopedia of Astronomers* (Heidelberg: Springer, 2009), 388.

3. American Institute of Physics, "Spectroscopy and the Birth of Astrophysics," Center for History of Physics, www.aip.org/history/cosmology/tools/tools-spectroscopy.htm.

4. F. Zwicky, "Die Rotverschiebung von extragalaktischen Nebeln" [The redshift of extragalactic nebulae], translated by Friedemann Brauer, *Helvetica Physica Acta* 6 (1933): 110–27. For these translations, I have relied on Sidney van den Bergh, "The Early History of Dark Matter," *Publications of the Astronomical Society of the Pacific* 111, no. 760 (June 1999): 657.

5. Sinclair Smith, "The Mass of the Virgo Cluster," *Astrophysical Journal* 83 (1936): 23–31.

6. Tricia Close, "Lunatic on a Mountain: Fritz Zwicky and the Early History of Dark Matter" (master's thesis, Saint Mary's University, Halifax, Nova Scotia, 2001), http://library2.smu.ca/bitstream/handle/01/22390/close_tricia_masters_2001.PDF.

7. "Kent Ford & Vera Rubin's Image Tube Spectrograph Named in Smithsonian's '101 Objects That Made America,'" Department of Terrestrial Magnetism, Carnegie Institution for Science, https://dtm.carnegiescience.edu/news/kent-ford-vera-rubins-image-tube-spectrograph-named-smithsonians-101-objects-made-america; and Derek J. de Solla Price, *Little Science, Big Science* (New York: Columbia University Press, 1963), 70.

8. Dark matter particles—whatever they may be—are expected to move at low speeds and are hence referred to as cold.

9. Jon Agar, *Science in the Twentieth Century—and Beyond* (Cambridge: Polity, 2012), 164.

10. Antoine de Saint-Exupéry, *The Little Prince,* translated by Katherine Woods (New York: Harcourt Brace and World, 1943), 48; and John F. Fulton, "Robert Boyle and His Influence on Thought in the Seventeenth Century," *Isis* 18, no. 1 (July 1932): 77–102.

11. Albert A. Michelson and Edward W. Morley, "On the Relative Motion of the Earth and of the Luminiferous Ether," *Sidereal Messenger* 6 (1887): 306–10. Michelson was awarded the Nobel Prize in Physics in 1907 for developing interferometers, optical precision instruments.

12. F. Zwicky, *Morphological Astronomy* (Berlin: Springer, 1957), 132. Although we can credit Zwicky with coining the term *dark matter,* J. R. Bond and A. S. Szalay introduced its modern usage referring to cold, collisionless particles into the literature in "The Collisionless Damping of Density

Fluctuations in an Expanding Universe," *Astrophysical Journal* 274 (1983): 443–68.

13. F. Zwicky, "On the Masses of Nebulae and of Clusters of Nebulae," *Astrophysical Journal* 86, no. 3 (1937): 237.

14. G. Soucail, "The Giant Luminous Arc in the Centre of the A 370 Cluster of Galaxies," *ESO Messenger* 48 (June 1987): 43–44, available at http://adsabs.harvard.edu/abs/1987Msngr..48...43S.

15. J. H. Oort, "Some Problems Concerning the Structure and Dynamics of the Galactic System and the Elliptical Nebulae NGC 3115 and 4494," *Astrophysical Journal* 91 (1940): 273.

16. F. D. Kahn and L. Woltjer, "Intergalactic Matter and the Galaxy," *Astrophysical Journal* 130 (1959): 705–17.

17. A. Wilson, "Zwicky: Humanist and Philosopher," *Engineering and Science* 37 (March–April 1974): 18.

18. Dennis Overbye, *Lonely Hearts of the Cosmos* (New York: Harper Collins, 1991), 18.

19. V. Rubin, K. Ford, and J. Rubin, "A Curious Distribution of Radial Velocities of ScI Galaxies with $14.0 \leq M \leq 15.0$," *Astrophysical Journal Letters* 183 (1973): L111.

20. Interview of Dr. Vera Cooper Rubin by David DeVorkin on May 9, 1996, Niels Bohr Library and Archives, American Institute of Physics, College Park, MD, www.aip.org/history-programs/niels-bohr-library/oral-histories/5920-2. Richard Panek describes Rubin's work and contributions in detail in *4% Universe,* 25–53; see also Alan Lightman and Roberta Brawer, *Origins: The Lives and Worlds of Modern Cosmologists* (Cambridge, MA: Harvard University Press, 1990), 291.

21. M. S. Roberts and R. N. Whitehurst, "The Rotation Curve and Geometry of M31 at Large Galactocentric Distances," *Astrophysical Journal* 201 (1975): 327–46.

22. J. Ostriker and J. P. E. Peebles, "A Numerical Study of the Stability of Flattened Galaxies: or, Can Cold Galaxies Survive?," *Astrophysical Journal* 186 (1973): 467.

23. J. Ostriker, J. P. E. Peebles, and A. Yahil, "The Size and Mass of Galaxies, and the Mass of the Universe," *Astrophysical Journal Letters* 193 (1974): L1; and Agris J. Kalnajs, "Halos and Disk Stability," in *Dark Matter in the Universe: Proceedings of the 117th Symposium of the International Astronomical Union,* ed. J. Kormendy and G. R. Knapp (Boston: Kluwer Academic, 1987), 289–99.

24. Virginia Trimble, "History of Dark Matter in Galaxies," in *Planets, Stars and Stellar Systems,* ed. Terry D. Oswalt, vol. 5, *Galactic Structure and Stellar Populations,* ed. Gerry Gilmore (New York: Springer, 2013), 1091–118.

25. George R. Blumenthal, S. M. Faber, Joel R. Primack, and Martin J. Rees, "Formation of Galaxies and Large-Scale Structure with Cold Dark Matter," *Nature* 311 (1984): 517; and S. D. M. White, C. S. Frenk, and M. Davis, "Clustering in a Neutrino-Dominated Universe," *Astrophysical Journal Letters* 274 (1983): L1; Vera Rubin, "The Rotation of Spiral Galaxies," *Science* 220 (June 24, 1983): 1344.

26. Thomas S. Kuhn, *The Structure of Scientific Revolutions* (Chicago: University of Chicago Press, 1962).

27. Richard Holmes, *The Age of Wonder: How the Romantic Generation Discovered the Beauty and Terror of Science* (New York: Vintage, 2010) 60–125.

28. A. Einstein, "Die Grundlage der allgemeinen Relativitätstheorie" [The foundation of the general theory of relativity], *Annalen der Physik* 49, no. 7 (1916): 769–822.

29. Arrigo Finzi, "On the Validity of Newton's Law at a Long Distance," communicated by F. A. E. Pirani, *Monthly Notices of the Royal Astronomical Society* 127 (1963): 28, 30.

30. Jacob Bekenstein and Mordehai Milgrom, "Does the Missing Mass Problem Signal the Breakdown of Newtonian Gravity?," *Astrophysical Journal* 286 (1984): 7–14; and Milgrom, "Does Dark Matter Really Exist?," *Scientific American,* August 2002, 42.

31. Jean-Paul Kneib and Priyamvada Natarajan, "Cluster Lenses," *Astronomy and Astrophysics Review* 19 (2011): article 47.

5. The Changing Scale

1. H. G. Wells, *The First Men in the Moon* (London: Newnes, 1901).

2. Roger Babson, "Gravity—Our Enemy Number One," reprinted in H. Collins, *Gravity's Shadow: The Search for Gravitational Waves* (Chicago: University of Chicago Press, 2010), 828–29; for the list of Gravity Research Foundation prizewinners, see www.gravityresearchfoundation.org/winners_year.html.

3. Richard Panek, *The 4% Universe: Dark Matter, Dark Energy, and the Race to Discover the Rest of Reality* (New York: Mariner Books, 2011), xv; this book has a well researched account of the search for supernovae and the discovery of dark energy.

4. Edward N. Zalta, ed., *The Stanford Encyclopedia of Philosophy,* s.v. "New-

ton's Philosophy," by Andrew Janiak, last revised May 6, 2014, http://plato.stanford.edu/archives/sum2014/entries/newton-philosophy/; and Panek, *4% Universe,* 60.

5. Panek, *4% Universe,* 60; and Isaac Newton to Richard Bentley, February 25, 1692, THEM00258, 189.R.4.47, fols. 7–8, Newton Project, Trinity College Library, Cambridge, U.K., accessed September 15, 2015, www.newtonproject.sussex.ac.uk/catalogue/record/THEM00258. Latin texts and English translations of the "General Scholium" from various editions of the *Principia* can be found at http://isaacnewton.ca/newtons-general-scholium/, part of the Newton Project Canada. See also Newton, *Philosophiae naturalis principia mathematica* (Cambridge: Cambridge University Press, 1687); and *Isaac Newton's Philosophiae naturalis principia mathematica: The Third Edition, 1726, with Variant Readings,* ed. A. Koyré and I. B. Cohen, with the assistance of A. Whitman (Cambridge, MA: Harvard University Press, 1972).

6. W. Baade, "The Absolute Photographic Magnitude of Supernovae," *Astrophysical Journal* 88 (1938): 285–304.

7. A. S. Eddington, *The Mathematical Theory of Relativity* (Cambridge: Cambridge University Press, 1923) 119–46, 152–61; and Eddington, *The Expanding Universe* (Cambridge: Cambridge University Press, 1933), 102.

8. A. Einstein and W. de Sitter, "On the Relation Between the Expansion and the Mean Density of the Universe," *Proceedings of the National Academy of Sciences* 18 (1932): 213.

9. For a more comprehensive treatment of all the measurable cosmological parameters, see Martin Rees, *Just Six Numbers: The Deep Forces That Shape the Universe* (New York: Basic Books, 2000), chapters 6–7.

10. Allan Sandage, "The Ability of the 200-Inch Telescope to Discriminate Between Selected World Models," *Astrophysical Journal* 133 (1961): 389.

11. For Kirshner's account of the discovery of dark energy, see his *The Extravagant Universe: Exploding Stars, Dark Energy and the Accelerating Cosmos* (Princeton: Princeton University Press, 2002), 158–262.

12. Panek, *4% Universe,* 71.

13. S. Perlmutter, G. Aldering, M. Della Valle, S. Deustua, R. S. Ellis, S. Fabbro, A. Fruchter, et al., "Discovery of a Supernova Explosion at Half the Age of the Universe," *Nature* 391 (1998): 51.

14. A. G. Riess, A. V. Filippenko, P. Challis, A. Clocchiatti, A. Diercks, P. M. Garnavich, R. L. Gilliland, et al., "Observational Evidence from Supernovae for an Accelerating Universe and a Cosmological Constant," *Astronomical Journal* 116, no. 3 (1998): 1009–38.

15. Panek, *4% Universe,* 158–59.

16. James Glanz, "Exploding Stars Point to a Universal Repulsive Force," *Science* 279, no. 5351 (January 30, 1998): 651–52, available at www.sciencemag.org/content/279/5351/651.summary?sid=c2d42164-3577-4952-9687-77cf531d4729.

17. Panek, *4% Universe,* 163; Marcia Bartusiak, *Archives of the Universe: 100 Discoveries That Transformed Our Understanding of the Cosmos* (New York: Vintage, 2004), 608–11.

18. Peter L. Galison, "Introduction: The Many Faces of Big Science," in *Big Science: The Growth of Large-Scale Research,* ed. Galison and Bruce Hevly (Stanford: Stanford University Press, 1992), 1; and W. K. H. Panofsky, "SLAC and Big Science: Stanford University," in ibid., 145.

6. The Next Wrinkle

1. For John Mather's account of the COBE launch, see Mather and John Boslough, *The Very First Light: The True Inside Story of the Scientific Journey Back to the Dawn of the Universe* (New York: Basic Books, 2008), 3–9; "satellite of love" is a reference to one of the singer Lou Reed's best-known songs, from his 1972 album *Transformer.*

2. Laurence Bergreen, *Over the Edge of the World: Magellan's Terrifying Circumnavigation of the Globe* (New York: Harper Collins, 2004); and Mather and Boslough, *Very First Light,* 255–63.

3. Derek J. de Solla Price, *Little Science, Big Science* (New York: Columbia University Press, 1963), 239; and Helge Kragh, *Cosmology and Controversy: The Historical Development of Two Theories of the Universe* (Princeton: Princeton University Press, 1996), 123–34.

4. G. Lemaître, "L'expansion de l'espace," *Revue des questions scientifiques* 20 (November 1931): 391–410, translated by Betty H. Korff and Serge A. Korff in *The Primeval Atom: An Essay on Cosmogony* (New York: Van Nostrand, 1950), quote on 78–79; and Lemaître, "The Beginning of the World from the Point of View of Quantum Theory," *Nature* 127, no. 3210 (1931): 706. See also Kragh, *Cosmology and Controversy,* 22–60.

5. Kragh, *Cosmology and Controversy,* 81–101; Mather and Boslough, *Very First Light,* 28.

6. For Ralph A. Alpher and Robert Herman's first results, see "Evolution of the Universe," *Nature* 162, no. 4124 (1948): 774–75. See also Mather and Boslough, *Very First Light,* 42–43.

7. P. J. E. Peebles, "Discovery of the Hot Big Bang: What Happened

in 1948," *European Physical Journal H* 39, no. 2 (2014): 205–23; Mather and Boslough, *Very First Light,* chapters 1–6, vividly recounts the details of the history of the discovery of the CMBR, including a thoughtful analysis of how and why Alpher, Herman, and Gamow's contributions went unrecognized (a take that I am in full agreement with).

8. Mather and Boslough, *Very First Light,* 44; and R. H. Dicke, R. Beringer, R. L. Kyle, and A. B. Vane, "Atmospheric Absorption Measurements with a Microwave Radiometer," *Physical Review* 70 (1946): 340–48.

9. Kragh, *Cosmology and Controversy,* 133–34.

10. George Gamow, *The Creation of the Universe* (New York: Dover Science Books, 1952; reissue, 2004), chapters 2–4.

11. Alan Lightman, *The Discoveries: Great Breakthroughs in 20th-Century Science, Including the Original Papers* (New York: Pantheon, 2005), 411.

12. R. H. Dicke, P. J. E. Peebles, P. G. Roll, and D. T. Wilkinson, "Cosmic Black-Body Radiation," *Astrophysical Journal* 142 (1965): 416.

13. Marcia Bartusiak, *Archives of the Universe: 100 Discoveries That Transformed Our Understanding of the Cosmos* (New York: Vintage, 2004), 508.

14. Interview of Ralph Alpher and Robert Herman by Martin Harwit on August 12, 1983, Niels Bohr Library and Archives, American Institute of Physics, College Park, MD, 78. www.aip.org/history-programs/niels-bohr-library/oral-histories/3014-2; and Mather and Boslough, *Very First Light,* 39–49, 61–62.

15. The President's National Medal of Science citation for Ralph A. Alpher, National Science Foundation, accessed August 25, 2014, www.nsf.gov/od/nms/recip_details.jsp?recip_id=5300000000427.

16. Details are presented in R. H. Dicke, "The Measurement of Thermal Radiation at Microwave Frequencies," *Review of Scientific Instruments* 17, no. 7 (1946): 268–75.

17. John Noble Wilford, "Scientists Report Profound Insight on How Time Began," *New York Times,* April 24, 1992; see also R. C. Smith, "Essay Review: The Significance of COBE," *Contemporary Physics* 35, no. 3 (1994): 213–15.

18. Peter Galison, *Image and Logic: A Material Culture of Microphysics* (Chicago: University of Chicago Press, 1997), 46, 830.

19. A longer account of the BICEP saga can be found in Priyamvada Natarajan and Ravi Sankrit, "Ringside Seat at the Cutting Edge of Science," *YaleGlobal Online,* June 12, 2014, http://yaleglobal.yale.edu/content/ringside-seat-cutting-edge-science.

7. The New Reality and the Quest for Other Worlds

1. The main mission of NASA's Kepler satellite, launched on March 6, 2009, is planet hunting. This is the first satellite designed and launched to do so. To date, Kepler has uncovered more than four thousand candidate exoplanets.

2. Martin Rees, *Just Six Numbers: The Deep Forces That Shape the Universe* (New York: Basic Books, 2000), chapters 1–2.

3. Ibid., chapter 3.

4. As all fans of Douglas Adams's cult classic *The Hitchhiker's Guide to the Galaxy* know, 42 is the answer to the Ultimate Question of Life, the Universe, and Everything.

5. W. R. Stoeger, G. F. R. Ellis, and U. Kirchner, "Multiverses and Cosmology: Philosophical Issues," preprint, last revised January 19, 2006, http://arxiv.org/abs/astro-ph/0407329.

6. Epicurus, "Letter to Herodotus," in *The Extant Remains,* translated by Cyril Bailey (Oxford: Clarendon, 1926), 45.

7. According to Bertrand Russell, Western philosophy begins with Thales. See *The History of Western Philosophy* (New York: Simon and Schuster, 1945), 15.

8. Ayman Shihadeh, "From al-Ghazali to al-Razi: 6th/12th Century Developments in Muslim Philosophical Theology," *Arabic Sciences and Philosophy* 15 (2005): 141–79.

9. J. T. Fraser, *Of Time, Passion, and Knowledge: Reflections on the Strategy of Existence* (Princeton: Princeton University Press, 1990), 32.

10. Bernard de Bovier de Fontenelle, *Conversations on the Plurality of Worlds,* translated by Elizabeth Gunning (London, 1803), 110.

11. Camille Flammarion, *Lumen,* translated by A.A.M. and R.M. (New York: Dodd, Mead, 1897), available at www.gutenberg.org/ebooks/43835; and Mark Brade, *Alien Life Imagined: Communicating the Science and Culture of Astrobiology* (Cambridge: Cambridge University Press, 2012), 194, 195.

12. Emily Swanson, "Alien Poll Finds Half of Americans Think Extraterrestrial Life Exists," *Huffington Post,* updated July 29, 2013, www.huffingtonpost.com/2013/06/21/alien-poll_n_3473852.html.

13. Steven J. Dick, *Life on Other Worlds: The 20th-Century Extraterrestrial Life Debate* (Cambridge: Cambridge University Press, 1998), 193–95; and George Gaylord Simpson, "The Nonprevalence of Humanoids," *Science* 143, no. 3608 (February 21, 1964): 769–75.

14. Dennis Overbye, "Far-Off Planets like the Earth Dot the Galaxy,"

New York Times, November 5, 2013; Doyle Rice, "Earthshaking News: There May Be Other Planets like Ours," *USA Today,* November 4, 2013; and Martin Rees, "Are We Alone in the Universe? We'll Know Soon," *Guardian,* September 16, 2012.

15. "Fermi Paradox," SETI Institute, www.seti.org/seti-institute/project/details/fermi-paradox; Michael Hart, "An Explanation for the Absence of Extraterrestrials on Earth," *Quarterly Journal of the Royal Astronomical Society* 16, no. 2 (1975): 135.

16. Stephen Jay Gould interview, Academy of Achievement, June 28, 1991, www.achievement.org/autodoc/page/gou0int-1; and Theodosius Dobzhansky, "Nothing in Biology Makes Sense Except in the Light of Evolution," *American Biology Teacher* 35, no. 3 (1973): 125–29.

17. The latest tally can be found at the Kepler page of NASA's Ames Research Center, http://kepler.nasa.gov/.

18. "Events," Breakthrough Initiatives, accessed September 22, 2015, www.breakthroughinitiatives.org/?controller=Page&action=page&page_id=6; and Rachel Feltman, "Stephen Hawking announces $100 million hunt for alien life," *Speaking of Science* (blog), *Washington Post,* July 20, 2015, www.washingtonpost.com/news/speaking-of-science/wp/2015/07/20/stephen-hawking-announces-100-million-hunt-for-alien-life/.

SUGGESTED FURTHER READING

There are many other books available to the curious reader who wants to follow up and learn more in detail about the cosmological ideas discussed here. This is a selection that covers the history of ideas in cosmology and surveys the current open scientific questions.

Ball, Philip. *Curiosity: How Science Became Interested in Everything*. Chicago: University of Chicago Press, 2013.

Barrow, John D. *The Book of Universes: Exploring the Limits of the Cosmos*. New York: W. W. Norton, 2012.

———. *The Constants of Nature*. London: Jonathon Cape, 2002.

Bartusiak, Marcia. *Archives of the Universe: A Treasury of Astronomy's Historic Works of Discovery*. New York: Pantheon, 2004.

———. *Black Hole: How an Idea Abandoned by Newtonians, Hated by Einstein, and Gambled On by Hawking Became Loved*. New Haven: Yale University Press, 2015.

———. *The Day We Found the Universe*. New York: Vintage, 2009.

Bernstein, Jeremy. *Albert Einstein: And the Frontiers of Physics*. Oxford: Oxford University Press, 1996.

Bronowski, Jacob. *The Common Sense of Science*. Cambridge, Mass: Harvard University Press, 1978.

———. *The Origins of Knowledge and Imagination*. New Haven: Yale University Press, 1978.

Brotton, Jerry. *A History of the World in Twelve Maps*. New York: Viking, 2012.

Carroll, Sean. *The Particle at the End of the Universe: How the Hunt for the Higgs Boson Leads Us to the Edge of a New World*. New York: Dutton, 2012.

Corfield, Richard. *Lives of the Planets: A Natural History of the Solar System*. New York: Basic Books, 2007.

Dalal, Ahmad. *Islam, Science, and the Challenge of History*. New Haven: Yale University Press, 2010.

Davies, Paul. *The Goldilocks Enigma: Why Is the Universe Just Right for Life?* New York: Allen Lane, 2006.

Davies, Paul, and J. Gribbin. *The Matter Myth: Dramatic Discoveries That Challenge Our Understanding of Physical Reality*. New York: Simon and Schuster, 2007 (reissue).

Dyson, Freeman. *The Scientist as Rebel*. New York: New York Review of Books, 2014.

Ferguson, Kitty. *Measuring the Universe: Our Historic Quest to Chart the Horizons of Space and Time*. New York: Walker Books, 1999.

———. *Tycho and Kepler*. New York: Walker Books, 2002.

Freese, Katherine. *The Cosmic Cocktail: Three Parts Dark Matter*. New York: W. W. Norton, 2003. Reprint, Princeton: Princeton University Press, 2014.

Galison, Peter L. *Big Science: The Growth of Large-Scale Research*. Stanford: Stanford University Press, 1994.

———. *Einstein's Clocks, Poincaré's Maps: Empires of Time*. New York: W. W. Norton, 2004.

———. *How Experiments End*. Chicago: University of Chicago Press, 1997.

Gates, Evalyn. *Einstein's Telescope: The Hunt for Dark Matter and Dark Energy in the Universe*. New York: W. W. Norton, 2009.

Gingerich, Owen. *The Book Nobody Read*. New York: Penguin, 2005.

Gleiser, Marcelo. *The Island of Knowledge: The Limits of Science and the Search for Meaning*. New York: Basic Books, 2014.

Goldberg, David. *The Universe in the Rearview Mirror: How Hidden Symmetries Shape Reality*. New York: Dutton: 2013.

Greene, Brian. *The Elegant Universe: Superstrings, Hidden Dimensions, and the Quest for the Ultimate Theory*. New York: W. W. Norton, 2003.

———. *The Fabric of the Cosmos: Space, Time, and the Texture of Reality*. New York: Knopf, 2004.

Gribbin, John. *Alone in the Universe: Why Our Planet Is Unique*. London: Wiley, 2011.

———. *In Search of the Big Bang*. London: Bantam, 1986.

———. *The Origins of the Future: Ten Questions for the Next Ten Years*. New Haven: Yale University Press, 2006.

Grinnell, Frederick. *Everyday Practice of Science: Where Intuition and Passion Meet Objectivity and Logic*. Oxford: Oxford University Press, 2009.

Hawking, Stephen. *A Brief History of Time: From the Big Bang to Black Holes*. London: Bantam, 1988.

———. *The Universe in a Nutshell*. London: Bantam, 2001.

Hellman, Hal. *Great Feuds in Science: Ten Disputes That Shaped the World*. New York: Barnes and Noble, 1998.

Holmes, Richard. *The Age of Wonder: How the Romantic Generation Discovered the Beauty and Terror of Science*. London: Pantheon, 2009.

Huth, John Edward. *The Lost Art of Finding Our Way*. Cambridge, MA: Harvard University Press, 2013.

Jaywardhana, Ray. *Strange New Worlds: The Search for Alien Planets and Life Beyond Our Solar System*. Princeton: Princeton University Press, 2011.

Kanas, Nick. *Star Maps: History, Artistry and Cartography*. London: Praxis, 2007.

Kirshner, Robert P. *The Extravagant Universe: Exploding Stars, Dark Energy, and the Accelerating Cosmos*. Princeton: Princeton University Press, 2004.

Kragh, Helge. *Conceptions of Cosmos: From Myths to the Accelerating Universe—A History of Cosmology*. Oxford: Oxford University Press, 2007.

———. *Cosmology and Controversy: The Historical Development of Two Theories of the Universe*. Princeton: Princeton University Press, 1996.

Krauss, Lawrence. *A Universe from Nothing: Why There Is Something Rather than Nothing*. New York: Atria Books, 2012.

Kuhn, Thomas S. *Essential Tension: Selected Studies in Scientific Tradition and Change*. Chicago: University of Chicago Press, 1977.

———. *The Structure of Scientific Revolutions*. Chicago: University of Chicago Press, 1962.

Levenson, Thomas. *Einstein in Berlin*. New York: Bantam, 2003.

Levin, Janna. *How the Universe Got Its Spots: Diary of a Finite Time in a Finite Space*. Princeton: Princeton University Press, 2002.

Liddle, Andrew, and Jon Loveday. *The Oxford Companion to Cosmology*. Oxford: Oxford University Press, 2008.

Lightman, Alan. *The Accidental Universe: The World You Thought You Knew*. New York: Corsair, 2013.

———. *The Discoveries: Great Breakthroughs in 20th-Century Science, Including the Original Papers*. New York: Pantheon, 2005.

———. *Einstein's Dreams*. New York: Pantheon, 1993.

Lightman, Alan, and Roberta Brawer. *Origins: The Lives and Worlds of Modern Cosmologists*. Cambridge, MA: Harvard University Press, 1990.

Livio, Mario. *The Accelerating Universe: Infinite Expansion, the Cosmological Constant, and the Beauty of the Cosmos*. New York: Wiley, 2000.

———. *Brilliant Blunders: From Darwin to Einstein—Colossal Mistakes by Great Scientists That Changed Our Understanding of Life and the Universe*. New York: Simon and Schuster, 2013.

Mather, John C., and John Boslough. *The Very First Light: The True Inside Story of the Scientific Journey Back to the Dawn of the Universe*. New York: Basic Books, 2008.

Mazlish, Bruce. *The Uncertain Sciences*. New Haven: Yale University Press, 1998.

Miller, Arthur I. *Empire of the Stars: Obsession, Friendship, and Betrayal in the Quest for Black Holes*. Boston: Houghton Mifflin, 2005.

Munitz, Milton K., ed. *Theories of the Universe: From Babylonian Myth to Modern Science*. New York: Free Press, 1957.

North, John. *Cosmos: An Illustrated History of Astronomy and Cosmology*. Chicago: University of Chicago Press, 2008.

Ohanian, Hans C. *Einstein's Mistakes: The Human Failings of Genius*. New York: W. W. Norton, 2008.

Ostriker, Jeremiah P., and Simon Mitton. *Heart of Darkness: Unraveling the Mysteries of the Invisible Universe*. Princeton: Princeton University Press, 2013.

Panek, Richard. *The 4% Universe: Dark Matter, Dark Energy, and the Race to Discover the Rest of Reality*. New York: Mariner Books, 2011.

Popper, Karl. *The Logic of Scientific Discovery*. 2nd ed. New York: Routledge, 2002.

Price, Derek J. de Solla. *Little Science, Big Science*. New York: Columbia University Press, 1963.

Primack, Joel R., and Nancy Ellen Abrams. *The View from the Center of the Universe: Discovering Our Extraordinary Place in the Cosmos*. New York: Riverhead, 2006.

Randall, Lisa. *Knocking on Heaven's Door: How Physics and Scientific Thinking Illuminate the Universe and the Modern World*. New York: Ecco, 2011.

Rees, Martin J. *Before the Beginning: Our Universe and Others*. New York: Perseus Books, 1997.

———. *Just Six Numbers: The Deep Forces That Shape the Universe*. New York: Basic Books, 2000.

———. *Our Cosmic Habitat*. London: Phoenix, 2002.

Scharf, Caleb. *The Copernicus Complex: Our Cosmic Significance in a Universe of Planets and Probabilities*. New York: Farrar, Strauss and Giroux, 2014.

———. *Gravity's Engines: How Bubble-Blowing Black Holes Rule Galaxies, Stars, and Life in the Cosmos*. New York: Farrar, Strauss and Giroux, 2012.

Shapin, Steven. *The Scientific Revolution*. Chicago: University of Chicago Press, 1996.

Shostak, Seth. *Confessions of an Alien Hunter: A Scientist's Search for Extraterrestrial Intelligence*. New York: National Geographic, 2009.

Silk, Joseph. *The Big Bang*. New York: W. H. Freeman, 2000.

———. *The Infinite Cosmos: Questions from the Frontiers of Cosmology*. Oxford: Oxford University Press, 2006.

Silvers, Robert B. *Hidden Histories of Science*. London: Granta, 1995.

Smolin, Lee. *The Life of the Cosmos*. Oxford: Oxford University Press, 1997.

———. *The Trouble with Physics: The Rise of String Theory, the Fall of a Science, and What Comes Next*. Cambridge, MA: Houghton Mifflin, 2006.

Sobel, Dava. *Galileo's Daughter: A Historical Memoir of Science, Faith, and Love*. New York: Walker Books, 2000.

———. *A More Perfect Heaven: How Copernicus Revolutionised the Cosmos*. New York: Walker Books, 2011.

Tegmark, Max. *Our Mathematical Universe: My Quest for the Ultimate Nature of Reality*. New York: Vintage, 2015.

Thorne, Kip S. *The Science of "Interstellar."* New York: W. W. Norton, 2014.

Tyson, Neil deGrasse. *Death by Black Hole: And Other Cosmic Quandries*. New York: W. W. Norton, 2007.

Tyson, Neil deGrasse, and Donald Goldsmith. *Origins: Fourteen Billion Years of Cosmic Evolution*. New York: W. W. Norton, 2004.

Wilford, John Noble. *The Mapmakers*. New York: Vintage, 2000.

ACKNOWLEDGMENTS

I owe the genesis for my longstanding curiosity about radical scientific ideas and their path to acceptance to people and places on three continents. I have been fortunate to be part of many unique and stimulating academic environments that have shaped my worldview as a physicist and as a writer. And my own life as a working scientist, generating, confronting, and testing new ideas challenges me to continually sharpen my thinking about the progress of radical ideas in modern astronomy.

The impetus to get started on this book began with Meg Jacob's enthusiastic encouragement. The Op-Ed Project at Yale and writing for *The New York Review of Books* provided me with opportunities to write for a broader readership. For this, I am immensely thankful to Katie Orenstein, Mark Lilla, and Bob Silvers. I am deeply grateful for the encouragement of friends from my many worlds—Yale University, the Radcliffe Institute for Advanced Study and 40 Concord, the Rockefeller Center at Bellagio, Trinity College, Cambridge, MIT, and Harvard. Over the years, conversations with Martin Rees, Evelyn Fox Keller, Amartya Sen, Russ Rymer, Richard Holmes, Geetanjali Singh Chanda, Nayan Chanda, Urmi Bhowmik, Susan Faludi, Bruce Mazlish, Owen Gingerich, Nancy Cott, David Kaiser, Rebecca Goldstein, Supratik Bose, John Huth, Alan Lightman, Brian Greene, Mario Livio, Gish Jen, Gail Mazur, Pilar Palacia, Judith Vishniac, and Peter Trachtenberg have inspired and influenced me. Three very loyal friends, Amy Barger, Wally Gilbert, and Jeremy Bernstein, were extremely generous with their time as I worked on this project. They assiduously read every word and offered exceptionally constructive comments. To them, my deepest thanks. Getting the book to its final form would not have been possible if it were not for Andrea Volpe—I cannot imagine how I would have managed without her expert guidance and constant cheering. Joseph Calamia and Jean Thomson Black, my editors at Yale University Press, offered wise counsel every step of the way. Working with the extremely efficient and enthusiastic team at Yale University Press—Sam Ostrowski, Juliana Froggatt, Margaret Otzel, James Johnson, Jennifer Doerr, and Maureen Noonan—has been a delight. Casey Reed expertly rendered the book's many diagrams and illustrations and designed an awesome cover. The Sterling Map Room and the Beinecke Rare Book & Manuscript Library at Yale were invaluable resources in both shaping the book and providing me with contemplative spaces to think and write. My family, as always, sustained me with their unconditional support.

INDEX

Note: Page numbers in *italics* refer to illustrations.